Successful Textbook Publishing

Books by Thomas D. Brock

Milestones in Microbiology, 1961, Prentice-Hall, Inc. reprinted 1975, American Society for Microbiology

Principles of Microbial Ecology, 1966, Prentice-Hall, Inc.

Biology of Microorganisms, 1st edition, 1970, Prentice-Hall, Inc.

Basic Microbiology with Applications (with K.M. Brock), 1st edition, 1973, Prentice-Hall, Inc.

Biology of Microorganisms, 2nd edition, 1974, Prentice-Hall, Inc.

Basic Microbiology with Applications (with K.M. Brock), 2nd edition, 1978, Prentice-Hall, Inc.

Thermophilic Microorganisms and Life at High Temperatures, 1978, Springer-Verlag

Biology of Microorganisms, 3rd edition, 1979, Prentice-Hall, Inc.

Membrane Filtration: A User's Guide and Reference Manual, 1983, Science Tech, Inc.

Biology of Microorganisms (with D.W. Smith and M.T. Madigan), 4th edition, 1984, Prentice-Hall, Inc.

General Microbiology: A Laboratory Manual (with Jane A. Phillips), 1984, Prentice-Hall, Inc.

Biotechnology: A Textbook of Industrial Microbiology (editor of the English-language edition), 1984, Sinauer Associates/Science Tech, Inc.

Successful Textbook Publishing: The Author's Guide, 1985, Science Tech, Inc.

Successful Textbook Publishing: The Author's Guide

Thomas D. Brock

Science Tech, Inc., Madison, Wisconsin

Thomas D. Brock
University of Wisconsin–Madison
1550 Linden Drive
Madison, Wisconsin 53706

Library of Congress Cataloging in Publication Data

Brock, Thomas D.
Successful textbook publishing.

Bibliography: p.
1. Text-books--Publication and distribution. 2. Text-books--Authorship. 3. Authors and publishers. I. Title.
Z286.T48B76 1985 070.5'2 84-27541
ISBN 0-910239-01-0

Science Tech, Inc., 1227 Dartmouth Rd.
Madison, WI 53705 USA

Printed in the United States of America
10 9 8 7 6 5 4 3 2 1

Preface

This book is for academic or nonacademic professionals interested in writing textbooks in their fields. It is a book about publishing, and about how the author must work with the publisher to bring out a successful book. Most people who will find this book useful are not writers but professionals active in their special fields of work. They will frequently be college professors or teachers, but nonacademic professionals such as scientists, engineers, physicians, humanists, and lawyers interested in writing textbooks should also find this book of value.

The textbooks dealt with are those used for instruction in colleges, universities, and other post-secondary institutions. Textbooks for elementary and high schools are generally not written by individual authors but by publishing teams and present quite different problems; they are not discussed here. The main thrust of the present book is commercial publishing, that is, the production of books that will be sold in the market place for use by students.

This book explains simply the author/publisher relationship. It discusses such important practical matters as contracts and royalties, copyright, putting together a manuscript that will be easy to turn into a book, how to work with editors, marketing and sales departments, the uses of computers and word processors to speed up book production.

This book is based on over 20 years experience writing textbooks. During that time, I have published 6 books, one of which has gone through two editions, another through four editions. Lately, I have become in a small way a publisher myself, so that I have practical experience on both sides of the author/publisher dichotomy. I hope that my unique experience as both author and publisher provides me with special insights into the publishing process that will be of value to both authors and publishers. But this book is written primarily for the author, and from the author's viewpoint.

A survey of professional writers showed that the average income in the early 1980's was about $5000 a year. However, textbook authors often make considerably more than the average, and textbook publishing is one of the most lucrative branches of the publishing industry. Whether a textbook is successful or not will depend primarily on writing and organizational abilities. But it can also depend, to a significant extent, on how knowledgeable the author is of the publishing process, and how able the author is to work with the publisher to produce a successful book.

I know of no other book like this one on the market. There are many style manuals and manuals of writing. Many of the major publishers have their own manuals for authors, but these tell the author primarily what the publisher wants the author to know. There are also many books on how to write or publish a trade book, but most of these are directed at writers of fiction or general nonfiction. The textbook author is unique in being a professional in a discipline first, and an author only second. I believe that the textbook author who reads this book first will save lots of time and money, and will end up with a more satisfactory product.

Probably nothing is more gratifying than publishing a textbook that finds wide use in the author's field. The chance to attract students to a field of study is really exciting, and the author who has produced a handsome book that is used by many students can find a large amount of personal satisfaction. I hope that in some way my book will help to make this possible, and I dedicate this book to all the students yet to come who may obtain satisfaction from their college textbooks.

Table of Contents

1

Introduction

A college textbook is defined as a book used by students as part of the instructional materials in a formal course in a college or university. The textbook should be contrasted with the trade book. Textbooks differ from trade books in a number of significant ways:

- Trade books are generally written by professionals making their living at writing, whereas textbooks are written by non-professional writers.
- Trade books are sold primarily through general book stores, whereas textbooks are sold through college and university book stores and are not usually available for sale in general book stores.
- Trade books are generally sold individually to buyers who decide to purchase based on an examination of the book itself, whereas textbooks are sold to students who have not themselves made the decision to purchase.
- Trade books are generally sold by the publisher to booksellers at large discounts, generally 40% or more, whereas textbooks are sold at short discounts of about 25%.
- Except for a relatively few ''blockbuster'' books, trade books are manufactured in relatively small numbers of copies whereas textbooks are often produced in much larger numbers.

- Whereas the purchaser of a trade book generally keeps the book after reading it, the purchaser of a textbook frequently sells the book back to the bookseller, who places it for resale. Thus, in the book store, the textbook is frequently competing with itself in the form of the used book.
- Partly because of competition with used copies, and partly because of advances in the field, textbooks must be frequently revised, whereas trade books may continue to sell without revision over many years.
- Developing and marketing costs are much higher for a textbook than for all but the most exceptional trade books.
- Textbook publishing is one of the most profitable forms of publishing, whereas trade book publishing is generally less profitable.
- Successful textbook authors frequently command large royalty payments, whereas the royalties to trade book authors, even for books with respectable sales, are generally lower.

The college textbook industry grew dramatically during the two decades 1960 through 1980, in parallel with unprecedented growth in college enrollments following the post-World War II baby boom. According to the Association of American Publishers, in 1960 total college textbook sales were an estimated $97 million, whereas by 1979 sales had grown to an estimated $825.6 million. It should be noted that these figures do not reflect total book sales, but only the sales of *new* books. Approximately 21% of total textbook sales to students consists of used books. Also, although total dollars have continued to rise at a rapid rate throughout the end of the decade of the 1970s, the number of new books sold increased at a lower rate, reflecting a leveling off in college enrollments. It can be expected that sometime in the 1980s there will be an actual decline in college textbook sales, reflecting the predicted decline in the number of full-time students.

A textbook is produced in quite a different way than a trade book. With trade books, the publisher generally does not actively develop each individual book, but works through independent authors agents or receives unsolicited manuscripts, some of which are published. With textbooks, on the other hand, the publisher usually actively cultivates authors, and plays a major role in the initiation and development of the idea for a book.

In publishing a textbook, the publisher and author provide contrasting abilities. The textbook author is generally ignorant of the publishing process but knows the academic field well. The publisher is usually quite ignorant of the discipline for which the textbook is being written but knows the publishing process well.

Publishing is a business. It deals with costs and expenses, payrolls, break-even analyses, marketing, advertising, and sales. Textbook authoring is not a business nor even a profession. For the textbook author, writing is a sideline, and whether or not the final book is successful, the author will still have a source of income.

When a textbook is written, the author's primary concern is that the field be covered well with a book which will be complete, attractive, and useful to students. The length of the book, the complexity of illustrative material, etc., are of secondary concern as the author tries to present the material in the best possible light. When the book is completed, the author considers it a success based on how well it does the job of presenting the material to students. On the other hand, when the publisher examines the completed manuscript of the book, the main concerns are quite different. The publisher looks first to see if the content of the book is such that it will fit into the market. If the main market is for a freshman-level book, and the author has produced a book that is intelligible only to juniors and seniors, then the market may be much smaller than the publisher wishes. Second, the publisher will ascertain how expensive the book will be to produce. If the price that the market will bear is $20 but the book in question is so expensive to produce that it can only be sold for $35, then this is a major concern. Only after the market and the cost of production have been considered will the publisher turn to the detailed content of the book itself, and consider how well the author has handled the material. Thus, the author is content-oriented whereas the publisher is product-oriented.

Since the publisher is not knowledgeable in the particular discipline, how are the market and the accuracy of the book evaluated? The publisher relies on the advice of those knowledgeable in the field. Every textbook publisher maintains a group of academic advisors, paid or unpaid, who examine proposed projects, outlines of books, and manuscripts, to see whether they will be suitable for the publisher's purposes. These advisors may initially be called on to suggest potential textbook authors. Subsequently, when a potential author has submitted an outline for a textbook project, the advisors will evaluate its worth. Finally, when the manuscript is completed, the advisors will look it over (hardly ever reading it word for word) to determine whether it has merit. The publisher's academic advisors may have some knowledge of the publishing business, but generally not much more than the author, so their viewpoint will be primarily that of the discipline for which the book is intended. Concurrently, the publisher's own staff will evaluate the proposal or the manuscript from the viewpoint of market and publishing economics.

It should be clear from what has just been said that finance has a major influence on the kinds of textbooks that are published. Although publishers can and do publish textbooks with small markets for more advanced courses, most publishers prefer to publish basic textbooks for the big freshman and sophomore markets. It is thus advisable that a potential author have both the intended market for the book and an understanding of the economics of publishing clearly in mind from the beginning of a project. Even if a contract has been signed, the publisher is not *bound* to publish the book if it is found to be unsuitable. Nothing can be more disheartening than for an author to spend a large amount of time writing a textbook and find that it is rejected because of market and production considerations.

The textbook business is peculiar because of the amount of leverage involved in the adoption of a book for a course. Although the student buys the book, the decision to purchase the book is made for the student by the professor teaching the course. Often (I am tempted to say generally) the professor has not read the book before deciding to adopt it. This puts the textbook author in a peculiar position. The book must be written so that it is *readable* by students, but it must be produced so that it is *attractive* to professors. A book that serves the student very well may be disastrous from a business viewpoint if professors do not like it. Thus, when developing and writing a textbook, the author must keep two masters in mind, the professor and the student.

Place of college textbook publishing in the United States publishing industry

Activity	Sales, 1981	Sales % of total industry	Net income, % of sales
Adult trade hardbound	$735.6	9.5	8.8
Adult trade paperbound	384.7	5.1	5.9
Juvenile	233.4	3.0	14.8
Religious	360.1	2.7	4.5
Professional	1140.7	2.6	14.6
Book clubs	571.1	1.8	9.5
Mail order	653.6	3.0	16.1
Mass market paperback	735.6	2.5	3.0
University presses	86.0	1.4	
Elementary and secondary texts	998.6	1.3	22.8
College texts	1074.7	2.5	22.5

Dollars are in millions. Data from the Association of American Publishers. Note that textbook publishing is the most profitable form of publishing, in terms of net income as percent of sales

Having said this, I must also confess that I have faith that despite the importance of impressing the professor, in the long run the successful book will be the one which serves the student best. Bad books which look good may find use for a year or two, but students will complain (often bitterly), and the book will quickly be dropped. The author should understand that the book is being written for the long haul, for periodic revision and updating, and for continued sales over many years.

I am convinced that it is possible to write a book that will appeal to both students and professors, and this should be the goal of all textbook authors.

2

Why write a textbook?

It often starts with contact by a publisher's representative. Say that for the past four years, you have been teaching a course for 100–500 students. Although the textbook you are using has problems (what textbook doesn't?), you had never considered writing one yourself. A publisher's representative from one of the major text houses drops by to see you. Their company lacks a book in your area and the book you are using is published by another company. You discuss the book and its defects and your course and its goals. The publisher's representative queries: "Have you ever considered writing a textbook yourself?" You hadn't, but are now urged to do so, and you initiate a dialog that continues over the next year or so. You talk with other publisher's representatives. The money mentioned sounds attractive. You ultimately sign a contract. You are writing a textbook.

Another route is possible. You initiate a new course. There is no textbook available that fits your course, so you use a detailed outline, notes, handouts, copies of articles from other sources. After teaching the course three or four times, you have refined the course so that you are satisfied with it. Although the course is successful, the students are unhappy without a formal textbook. You decide to write your own textbook. You put together your course material and contact a publisher. Although your course is new, the publisher sees that it is the forerunner of many others and expresses interest in doing your book. You sign a contract and begin writing.

A third route is possible. You have been teaching that large-enrollment course for some years and it has become your niche in the department. You have used a variety of textbooks, and although they have been generally satisfactory, you see nothing special about them. You have heard that textbook writing is a profitable side-line, and you could use the extra money. If one of your colleagues can write a textbook, why not you? Contact with a publisher's representative leads to a contract and you begin writing.

Another route is more subtle but still possible. You haven't actually taught the course very much, but have broad knowledge of the field. You are a good writer, a good organizer, perhaps known widely in your field. You find the textbooks in your field out-of-date, poorly written, badly produced. You feel an obligation to make your field interesting and attractive to students. Perhaps you have already written a monograph or advanced textbook so that you have some idea of how books are put together. You decide that you will do that much needed textbook. Because of your reputation, you have no trouble finding a publisher, and sign a contract without having written a single word. Now comes the difficult job of putting into words your vaguely formulated ideas.

What are your goals in textbook writing? Writing a textbook entails a large commitment of time and energy. It is not an enterprise to be entered into lightly. If you are to put your sweat and tears into such a project, you should have some clearly defined goals for its outcome.

One of the commonest results of publishing a textbook is professional advancement in your field. To write the textbook, you have had to survey the whole discipline. You have obtained new insights for your own scholarly work. You may even have seen new areas of scholarly activity that no one was working in, or have obtained ideas for scholarly studies that are unique. In initiating these new scholarly studies, you may make significant advances in the field which can result in professional recognition by peers.

When your book finds approval by students, and achieves the educational objective that you have sought, a large degree of personal satisfaction develops. Certainly there is no more tangible evidence of achievement than a handsome book with your name on it. It is not so much that you are seeing your name in print, but that the book is concrete evidence of your accomplishment.

Although no one in academic work is motivated primarily by the desire for fame, this may be one result of a successful book. The name of a textbook writer will be spread widely across the discipline, through advertisements, mailers, exhibits at professional conventions, discussion with colleagues by publisher's representatives. If you have done a good job on the book, people will know you and will respect you. You will achieve a kind of recognition in your field that is difficult to achieve through scholarly activities alone. (There are some negative aspects to this recognition; see Chapter 13.)

Last but not least, writing a textbook can bring money to the author. It is especially nice to consider the extra income because the writing of the textbook has not been done for the sake of the money itself. Not being a professional writer, the textbook author receives other rewards from the writing of the book, so that if the book is professionally and educationally sound but (heaven forbid) not successful financially, all is not lost. However, the amount of time and nervous energy involved in writing the book is enormous, and one would hope that some financial rewards result. We will discuss money in some detail in Chapter 11, but note here that the income from a textbook will depend not only on how successful the book is, but the size of the market for the book. A textbook written for an advanced course will not generate as much income as one written for an elementary course. (However, the book written for the advanced course is more likely to help its author professionally.) Whatever the other benefits, one hopes that enough money is generated from writing the book to pay off the mortgage on the house, or to send the children through college.

What are the first steps in developing the concept for a textbook? Generally, textbooks are written with existing courses in mind. A point of caution: Every academic course represents a compromise to local conditions. One must begin with the idea that the book is not just for the local course, but must find wide use. The local course materials: outline, notes, handouts, illustrative material, etc., may transfer perfectly into a textbook, but on the other hand they may well not. One must begin by stepping back from the local course and examining the field objectively. Colleges and universities vary enormously in how they are organized, how many credits they give for a certain amount of material, the length of the term, the number of lectures per week, etc. The successful textbook will be one that has a broad market.

Thus, the first thing the author must do is generalize the course. It

The wages and burden of textbook writing

A. Royalty income from textbook sales anticipated over the life of an edition (4-5 years)

Number of sales	List price			
	$15	$20	$25	$30
5,000	$ 7,500	10,000	12,500	15,000
10,000	10,000	20,000	25,000	30,000
20,000	30,000	40,000	50,000	60,000
50,000	75,000	100,000	125,000	150,000
100,000	150,000	200,000	250,000	300,000

Based on royalty as 10% of list. Other ways of calculating royalty are discussed in Chapter 11. Respectable sales for a basic college text would be in the 20,000–50,000 copy range. For an intermediate or advanced text, respectable sales would be in the 3,000–5,000 copy range.

B. The time needed by the author to complete a textbook

Activity	Basis	Total hours
Library research	1 year at 2 hr/week	104
Writing the book	2 years at 2 hr/day, 6 days/wk	1248
Assembling illustrative material	12 weeks at 2 hr/day, 6 days/wk	144
Production activities, copy editing, proof reading, checking art, indexing, etc.	1 year at 2 hr/day, 6 days/wk	624
Total hours		2120

The library research is that specifically related to the textbook, not research that the author does for scholarly activity. It is assumed that the author has a solid knowledge of the subject prior to the initiation of the textbook project.

C. Hourly rate for preparing the book, based on 2120 total hours and anticipated royalty

Royalty	Hourly rate
$ 7,500	$ 3.50/hr
10,000	4.72
50,000	23
100,000	47
300,000	141

is very important, at this early stage, to work as much as possible with the publisher. Everything becomes easier and more efficient if the author and publisher develop the concept for the book first, before any significant writing has been accomplished. We will have much to say about this in subsequent chapters.

The publisher approaches this whole enterprise much differently than the author. There is a high-enrollment course for which the publisher lacks a textbook. A survey by the publisher's field representatives or by an independent organization indicates that there is a market for a new textbook in that area. Individuals in the publisher's organization, called "acquisitions editors", are responsible for developing new textbooks (sometimes called "properties" or "projects"). A search for a suitable author develops. Academic advisors are queried. Discussion of the project may occur at regional or national sales meetings, where acquisitions editors and field representatives get together. Several potential authors are identified, and the acquisitions editor telephones or visits these to see if there is any interest. Many contacts are made at professional conventions in the particular discipline. There are many potential authors out there, but the acquisitions editor is looking for the "right" author.

Who is the "right" author? This is frequently the individual who teaches the large-enrollment course at a respected institution, is well known to colleagues, and has done some book writing before. Such an individual is often very busy, and frequently uninterested in writing the proposed book. Thus, the problem becomes one of finding the right "interested" author, a much more difficult matter.

We thus see that the aims of the publisher and the aims of the author are really quite different at the beginning of the project. The author sees the book in terms of the *course* and discipline, whereas the publisher sees the book in terms of the *market*. The main problem in publishing a successful textbook is bringing the author's goals and the publisher's goals together. The author must overcome any special foibles and strive to ensure that the book is done in such a way that it has a wide market. On the other hand, the publisher must resist the temptation to be too "commercial" and must strive to ensure that the book is done in such a way that it is academically sound. Much of the difficulty that occurs during the development of a textbook project arises because of inability of author and publisher to see each other's points of view.

What qualities does an author need to write a successful textbook? Here are the qualities that I think are essential for the writing of a successful textbook.

- Knowledge of the field
- Organizational ability
- Ability to see students viewpoint
- Ability to see publisher's viewpoint
- Ability to take criticism, often frank or impolite
- Ability to write reasonably well

Signing a contract with a publisher is such an important thing that I have reserved a whole chapter for it (Chapter 9). At this point, the main thing that needs to be said is: **Don't sign the contract without reading this book! The publisher's printed contract is just the starting point for negotiations.** These negotiations are often complex and detailed, and may, at some stage, even involve your lawyer. Signing a publishing contract, especially for a major college textbook, is not unlike taking marriage vows. You are about to become wedded to the publisher for life! You are giving away many things when you sign the contract. The publisher is also giving away some things, but fewer. (This is a result of the inequivalence of the Author/Publisher relationship.) Take it easy!

Copyright in your work is a matter which is also complex and frequently misunderstood. I have reserved a whole chapter for copyright (Chapter 10), to cover the many complexities involved. It is important to understand that even though the work you have written is copyrighted in your name, when you sign the contract you transfer the copyright to the publisher, generally for life. A copyright can be viewed as something like a property right. You can build and own a house, but you can also sell this house to someone else, who then has complete right to do anything with this house, even tear it down. Once an author has signed a contract, all rights to the work developed under this contract are transferred to the publisher, who can do anything with these rights, even nothing. Copyright matters are important and complex, and must be considered in relation to the contract.

A college textbook begins as an author's baby, but ends up as a publisher's ward. Just as a parent loses control of a child when it grows up, so does an author lose control of a book when it is finished. Fortunately, the goals of the publisher generally coincide with the goals of the author, and the baby is not abused. The more the author understands the publishing process, the better able the author will be to influence the publisher's activities regarding the production and marketing of the author's book.

3

Who are the textbook publishers?

The publishing industry is large, and college textbook publishing is only a small part of its activity. The Association of American Publishers (AAP, the trade organization for the major U.S. publishers) separates the publishing industry into a number of basic categories: Trade; Religious; College Text; Elementary and Secondary Text; Professional (Scholarly and Monograph); University Presses. There are publishers that fit neatly into one or another of these categories, but they are the exception rather than the rule. Most publishers market books in more than one category, and some of the larger publishers have activities in virtually all of the categories. Even University Presses do not always fit their category perfectly, because many of them publish textbooks and trade books as well as their more traditional scholarly fare.

The publishing industry seems always to be in a state of change. Mergers and acquisitions may lead to the demise of established publishers, whereas new companies are always starting. Some of the larger publishers have acquired so many smaller companies that they have different divisions that are competing with themselves. The author who is entering this jungle for the first time will find it difficult to sort things out, especially if primary contacts are with only one or a few field representatives.

The AAP College Textbook Publishers

The major college textbook publishers are members of the College Division of the AAP. A recent listing is displayed in the Appendix. Prentice-Hall and McGraw-Hill were ranked in first and second positions among the top 12 college publishers in terms of 1983 revenue, far outranking their nearest rivals. Although not all college textbook publishers are members of the AAP College Division, the major ones are. Membership in the AAP College Division does not guarantee quality publishing, of course, but an author should be especially cautious of signing with a publisher who is *not* a member. Membership in the AAP at least is a suggestion of corporate stability.

Useful tools for learning about textbook publishers

There is a large amount of information available about the publishing industry, and every college or university reference library will have some useful books that the potential textbook author should examine.

An excellent analysis of the culture and commerce of the publishing industry has been written by L.A. Coser, C. Kadushin, and W.W. Powell (see bibliography). This book is required reading for any potential author who really wants to understand what he or she is getting into when signing a contract. Although the book deals with all aspects of publishing, textbook publishing is given fairly detailed treatment, and its extensive index can be used to locate topics of particular interest. A good discussion is given of a typical textbook publishing house, showing how textbook publishing differs from some of the other branches of the publishing business.

Another useful resource for the potential author is the Literary Market Place (LMP), a book published annually by R.R. Bowker Company, which gives names and addresses of organizations involved in all aspects of the publishing business. The most important section of LMP is that which lists the names, addresses, officers, organization, and output of every United States publisher that has produced more than 3 new titles in a given year. For each publisher, the size of the output, the number of new titles, and their areas of specialty are given. For the publishers that are members of conglomerates, the parent organization is given, as well as all of the other subsidiaries and imprints. This section of LMP is well indexed, and publishers are listed by major categories (trade, text, etc.) as well as by a number of minor categories such as encylopedias and dictionaries, juvenile, maps and atlases, art books, and subscription and mail order books. About 200 separate publishers of college textbooks are listed. Many of these are very small publishers specializing in particular fields, and are not members of the AAP College Division. To use this index more intelligently, the potential textbook author should also refer to an adjacent index which breaks down publishers by subject matter. Here are listed the names of publishers specializing in such areas as anthropology, behavioral sciences, biological sciences, business, engineering, education, govenment, history, communications, language arts, law, medicine, mathematics, physical sciences, social sciences, womens's studies, etc. By cross referencing between these two indexes, the textbook author can learn the names of the publishers specializing in any particular field.

Once the names of publishers have been selected, it is then desirable to look over their current lists of in-print books to see what kinds of things they have published. Most of the major publishers have their catalogs published yearly in another Bowker book called Publishers Trade List Annual. Although this emphasizes trade books, textbooks will also be listed if a publisher also has those on its list. For publishers not listed in the Publishers Trade List Annual, it should be possible to obtain a recent catalog from the textbook department of the campus bookstore. One can also obtain a catalog from any publisher by writing to its sales office (addresses in LMP).

Another useful tool for currently available books is the Bowker publication Books In Print. This multi-volume work, published annually, will be available in any library and in most bookstores. A book is "in print" when it is still available in one or another edition from a publisher. Books In Print lists by title and author and there is also a subject index, which may be useful for locating books in a particular field. Books In Print is "book-oriented" rather than "publisher-oriented", and it is useful in determining whether any books competing with the intended one already exist.

If an author is trying to select a publisher from a small list, it is strongly recommended that the general catalog for each publisher on the list be obtained, so that the kinds of books the publisher has already published in a given area can be determined. Specifically, are there competing books already on the publisher's list? Are there related books on the list that are used in more advanced courses? What are the prices? Is there any evidence that the publisher knows how to publish and market books in the particular field of interest?

The next step would be to examine copies of books brought out recently by each publisher on the final list. These might be seen in the college bookstore, or provided by the publisher's field representative. Look over recent editions carefully for evidence of quality production. We will discuss important aspects of the production process in later chapters.

Large and small textbook publishers

The publishing industry is complex. Some of the largest publishers publish books in all categories. Smaller publishers tend to be more specialized, but may still publish books in a variety of fields. There are some major economies of scale involved in the publishing business, which pushes publishers toward becoming larger and larger.

The larger companies publish textbooks in virtually every field. Some are so complex that they publish textbooks in various divisions that compete among themselves, leading to a situation that is not very satisfactory for the author. The sales organizations in these large companies frequently specialize, so that there may be more than one field representative calling at a given college or university. In the larger companies, sales are made primarily by face-to-face contact between the field representative and the professor adopting the book. The field staff in these companies may number over a hundred, and is in a continual state of flux, as trainees move into and up the ladder, and oldtimers move on to the main office or to other companies.

In smaller companies, the field staff may be small or even nonexistent. In these companies, books are sold primarily by direct mail advertising. Copies are sent to those professors answering response cards provided as part of the advertising, and adoptions occur because of the reputation of the author or the appearance of the book.

One point that needs to be emphasized is that publishers are usually not printers of books, and generally carry out very little if any of the book manufacturing process. Each publisher contracts with printers for the actual manufacture of books, and may work with many different printers. In the same way, printers work with many different publishers. Thus, from a book manufacturing standpoint, there is *no* advantage of a large over a small publisher. As we will discuss in detail in Chapter 6, virtually all significant aspects of book manufacture are performed outside the publisher's organization, by free-lance workers or by other companies. The small publisher has essentially the same access to these outside organizations as the large publisher. Where small and large publishers differ significantly is in sales and marketing activities. The large publisher markets books primarily through field representatives, only secondarily through direct mail advertising, and virtually not at all through advertisements in journals. The small publisher, unable to support a field staff, must rely on direct mail and journal advertising.

This does not mean that the small publisher cannot market textbooks as effectively as the larger publisher. On the contrary, in a specialized area the small publisher may be more effective. The marketing department and field representatives of the large publisher may have 100 new books or new editions to sell in any given year, and can obviously give careful attention to only a few. The small publisher, with only a few books to market in any given year, can carefully consider each book and develop a marketing plan that is most effective for that book. The small publisher may actually do better with certain kinds of books than the large publisher, although either kind of publisher may do quite well (or badly) with any given book.

Some major textbook publishers

Because of economies of scale (primarily in sales and marketing), and because of the nature of corporate activities, the larger publishers tend to be more profitable than the smaller publishers. Because they are more profitable, they are able to absorb smaller companies, and hence tend to become progressively larger. At times, over 40% of all textbook sales have been made by four companies, Prentice-Hall, McGraw-Hill, CBS Educational and Professional Publishing, and Scott-Foresman (the latter publishes primarily in the school textbook market, however). I have emphasized above the changing nature of the publishing industry. The index of the latest edition of LMP should be examined for a list of the current textbook publishers. A publisher that might have done a good job with a particular book last year may have changed so much that it would be highly inadvisable to sign with it this year. In the rest of this section, I give a brief analysis of the major textbook publishers, as determined by an analysis of LMP and from my own experience.

Prentice-Hall

Founded in 1913, this huge publisher brings out over 1000 new titles or new editions each year. It is one of the major college textbook publishers, but also publishes in virtually every other field. It has the following divisions: General Publishing Division (trade books); Children's Book Division; College Books Division (textbooks); Special Publications Division; Business and Professional Books Division; Education Books Division (school textbooks); Executive Reports Division (reference guides and newsletters for business); and a division which provides tax and law services. It includes a number of subsidiary companies that publish under their own imprints but come under the overall corporate umbrella. It also has separate divisions such as Appleton-Century-Crofts, a venerable publishing house acquired by Prentice-Hall in the early 1970s which now publishes primarily in the medical and nursing fields. In addition to its corporate headquarters in Englewood Cliffs, New Jersey, Prentice-Hall has a strong international operation, regional offices throughout the country, and field representatives seemingly in any location where college textbook users might exist.

McGraw-Hill

Whereas Prentice-Hall sprawls across the New Jersey bluffs, McGraw-Hill towers over 6th Avenue in New York City. Although perhaps Prentice-Hall's main competitor, McGraw-Hill is quite a different company. Founded in 1909, it publishes almost 1000 new titles a year, but also has many publishing activities outside of books. It is a major publisher of magazines and professional journals (Business Week is one of its best-known magazines), and through acquisition has acquired a number of smaller publishing companies which operate essentially independently. Although it publishes trade books, this sort of publishing is not its strong point. McGraw-Hill is divided into a number of divisions, of which the College Division is responsible for textbooks. Other divisions are the Health Professions Division, Professional and Reference Books Division, General Books Division, School Division, and Gregg Division. Among its subsidiaries is Osborne/McGraw-Hill, which publishes computer books, and Shepard's/McGraw-Hill, which publishes law books. McGraw-Hill also has strong international activities through McGraw-Hill International Book Co. A major factor on any college campus, if the field is science, business, or technology, McGraw-Hill almost certainly has a strong textbook position.

CBS Educational and Professional Publishing

This division of the large CBS communications empire consists of several book publishers which have been acquired by merger. Three major textbook publishers that are now under the CBS umbrella are Holt, Rinehart and Winston, The Dryden Press, and W.B. Saunders. Saunders has responsibility for science and medicine, Dryden for economics and business, while Holt retains responsibility for humanities and other fields. Whereas Holt has its headquarters on Fifth Ave. in New York, Saunders is located in Philadelphia. One of the examples of how corporate mergers and restructuring can affect authors can be illustrated by the example of CBS. Before CBS acquired Saunders, Holt was a general college textbook publisher, with a respected line of books in the science fields. When Saunders was acquired, CBS reshuffled the book lines, moving all science books to Saunders in Philadelphia. Authors who had signed with Holt in New York now found that their books were being edited by Saunders in Philadelphia. This is not to say that Saunders cannot do a good job of editing a science book, but it does indicate how authors (and indeed, all employees of a publishing company) can end up as pawns in a corporate merger situation. Other companies owned by CBS include the trade book publisher Praeger (New York City) and Winston Press, a Minneapolis publisher of preschool and school books.

Scott, Foresman

One of the largest textbook publishers, Scott, Foresman is best known for its books at the elementary and secondary school level, but it also publishes at the college level, primarily for the freshman and sophomore markets. This publisher has made a major investment in the development of microcomputer software for the education market, and is also a publisher of testing materials for schools. Founded in 1896, the company is located in Glenview, Illinois, and publishes around 500 new titles a year.

Harper and Row

Located in New York, Harper and Row is a major text and trade book publisher in its own right, as well as through one of its subsidiaries, J.B. Lippincott of Philadelphia. Founded in 1817, Harper and Row has published as many as 1000 new titles and new editions a year. Its major publishing divisions are: Trade Division, Junior Books, College Division, School Division, and International Division. Harper distributes books for a large number of other publishers, and in addition to Lippincott, has a number of other subsidiaries and imprints, including: Basic Books, Abelard-Schuman, Ballinger, Barnes and Noble, Thomas Y. Crowell, Funk and Wagnalls.

Other large publishers

In addition to the above major publishers of college textbooks, there exists also an important group of medium-sized publishers which have large college textbook divisions and publish in most of the major academic disciplines.

Addison-Wesley

This Massachusets publisher, founded in 1942, has a large textbook publishing program as well as activities in trade books and other fields. Its subsidiary is the California-based textbook publisher Benjamin/Cummings. Addison-Wesley is organized into the following divisions: Higher Education Publishing Group, World Science Division, World Language Division, General Books Division, School Publishing Group. The Higher Education Publishing Group has offices in both Massachusets and California, the latter together with Benjamin/Cummings.

Random-House

Although primarily known as a trade book publisher, this New York-based company also has extensive activities in the textbook area. Founded in 1925, it has expanded through growth and acquisition until it now publishes around 500 new titles a year. One of its subsidiaries is the respected trade book publisher Alfred A. Knopf Inc. Random-House is organized into the following divisions: Adult Books, Juvenile Books, Reference Division, School Division, College Division, International.

Academic Press

Founded in 1942, this company is a subsidiary of the trade book publisher Harcourt Brace Jovanovich. It has extensive publishing activites in the medical, scientific, and technical fields. Academic Press is primarily a publisher of monographs and symposium proceedings, but does publish some textbook volumes. Although based in New York throughout most of its existence, much of the Academic Press operations were uprooted and transferred to San Diego and/or Orlando when its parent company moved.

D.C. Heath

Best known for its books in the school area, this Massachusetts-based publisher also publishes books at the college level, primarily for the freshman and sophomore markets. Founded in 1885, the company publishes around 200 new titles a year.

Houghton Mifflin

This Boston-based publisher was founded in 1832 and publishes around 500 new titles a year. A major textbook publisher with strong titles in other lines, it is organized into the following divisions: Trade Book Division, Children's Books, Reference Division, International Division, School Division, College Division. In college textbook publishing, Houghton Mifflin has extensive activities in humanities, social sciences, science, engineering, business, economics, and mathematics.

W.B. Saunders

See CBS Publishing

Macmillan Publishing Co.

This New York-based company is a subsidiary of the British publisher Macmillan Inc. It also has its own subsidiaries: Collier, Crowell-Collier, Berlitz Publications, The Free Press, Hafner Press, and Schirmer Books (among others). Although it has extensive trade book offerings, Macmillan is perhaps better known for its school and college textbooks. It has the following divisions: School Division, College Division, Professional Books Division, General Books Division.

John Wiley and Sons

Founded in 1807, this New York-based company is one of the largest publishers of professional and technical books, but also has some textbook operations. John Wiley publishes around 800 new titles a year, and has extensive international offices which operate more or less independently of the parent company. In the college textbook field, it is noted for its books for advanced courses, primarily in science, engineering, and mathematics. It has the following divisions: Educational Group (college textbooks), Professional Group, Wiley Law Publications, Wiley Medical, International Group. Subsidiaries of John Wiley include Halsted Press, Heyden, Ronald Press, and Interscience.

Little, Brown

This Boston-based publisher, best known for its strong trade book line and its publication of the magazine Atlantic Monthly, also has a significant position in college textbooks and it is a major publisher of textbooks for the medical field. Founded in 1837, it publishes around 300 new titles a year. Its college textbook operations are now located primarily in California, whereas its trade and medical publishing is done from Boston.

Smaller significant publishers

Below are mentioned a number of other publishers that are not as large or as well known in the college textbook field, but which maintain textbook operations in at least some of the academic fields.

Charles E. Merrill

This Ohio-based company is a subsidiary of Bell and Howell Company, and is noted primarily for its textbooks at the elementary and high school levels. It is also a major publisher of tests and testing materials. Founded in 1842, it publishes around 200 new titles a year.

C.V. Mosby

This St. Louis-based publisher, a subsidiary of the Times Mirror Co., specializes in the publication of textbooks in the medical and biomedical areas. Founded in 1906, it publishes around 150 new titles a year and is also a publisher of journals in the medical field.

W.W. Norton

Although primarily known as a trade book publisher, this New York-based company has significant activities in college textbook publishing. Founded in 1924, it publishes around 200 new titles per year.

St. Martin's Press

A major trade book publisher, this company publishes over 500 new titles a year. In addition to trade books, it also has divisions for publishing college textbooks and reference and scholarly books.

Wadsworth

This California company, a subsidiary of the International Thomson Organisation, publishes around 200 new titles a year, primarily in the education field.

Bobbs-Merrill

This Indianapolis-based company is a subsidiary of Howard Sams Publishers. It publishes trade books from New York, law books from a subsidiary in Virginia, and textbooks from Indianapolis. Although known primarily for its elementary and high school texts, it also has some publishing activities at the college level.

Oxford University Press

Although primarily a publisher of scholarly books, this very old British company (founded in 1478!) also has extensive textbook activities. It publishes around 800 new titles a year.

Cambridge University Press

Slightly smaller and somewhat younger than Oxford (Cambridge was not founded until 1521), this British company publishes around 500 new titles a year, including many in the textbook field. Both Oxford and Cambridge are atypical university presses, since they have extensive trade book divisions and very active acquisitions programs.

Specialist publishers

As noted earlier, good textbooks are not published soley by large publishers. A number of quite small publishers have carved out niches for themselves in this field, doing careful editing and production jobs on a few books a year. Many of these publishers specialize in certain areas of college textbook publishing. Space does not permit a listing of any but the largest or most interesting, and the Literary Market Place indexes should be consulted for others.

W.H. Freeman

This New York/San Francisco-based company specializes in the production of reference books and textbooks in science and technology. Founded in 1946, it has well-respected books in a number of the sciences, and publishes around 40-50 new titles a year. In 1982, Freeman underwent a major reorganization and most of its significant publishing activities were moved from San Francisco to New York.

Wm. C. Brown

Located in Dubuque, Iowa, this publisher is known for field guides and other books used in biology teaching. Founded in 1944, it publishes around 100 new titles a year.

Sinauer Associates

This interesting publisher, based in the crossroads community of Sunderland, Massachusetts, has developed a strong list of textbooks in biology. It specializes in textbooks for the more advanced course, junior/senior level, where production costs are lower, and takes great care to produce high quality books. It markets virtually exclusively by direct mail, sending approval copies to professors returning response cards.

Worth Publishers

So small that it is not even listed in Literary Market Place, Worth is an example of a company dedicated to producing a few major textbooks well. Its emphasis is on the "blockbuster" textbook, one that sells to the large introductory market. It has successfully published major textbooks in biochemistry, biology, and botany which have become the leaders in their respective fields.

Final note

I have tried to provide in this chapter a feel for how rapidly the publishing industry can change. The reader should assume that anything written above is subject to unpredictable change. Acquisitions, mergers, bankruptcies, or new company formations can and will occur. The information presented in this chapter is intended merely to give an idea of the diversity and flavor of college textbook publishers. The latest edition of Literary Market Place and the latest directory of college publishers from the AAP should be consulted for current information.

4

Overview of the textbook publishing process

There are a number of steps in the publication process for any book. Publishers vary in exactly how they organize and implement these steps, but the overall process is fairly similar for all publishers. These are the main steps in publishing:

- Acquisitions
- Editorial and production
- Manufacturing
- Marketing
- Sales
- Order fulfillment

Within each of these broad categories, many subcategories are possible. In this chapter, we will discuss each of these major steps in the publishing process, how it is handled by textbook publishers, and how it affects the textbook author. By understanding what is going on "behind the scenes" at the publisher, the author will be in a better position to help guide the book through to a successful conclusion.

Many of the jobs in a publishing company have the word "editor" in the title, such as "acquisitions editor", "production editor", "managing editor'', ''copyeditor'', ''art editor''. In general, editors are those who are involved in some way with the actual production of the physical book. One major distinction in many publishing houses occurs between the editorial departments and the marketing and sales departments. Marketing and sales people are never called editors, and do not think of themselves as editors. Further, an individual who was an editor but has moved up the corporate ladder into management will drop the ''editor'' title. The textbook author will mostly deal with people in editorial departments, and only briefly with sales and marketing personnel.

Acquisitions: the key to successful publishing

The person with whom the textbook author will have the most contact is the acquisitions editor. The duties of an acquisitions editor include finding authors and working with them on the development of manuscripts up to the point where production of the book can begin. Other titles sometimes used are: procurement editor, development editor, commissioning editor, or sponsoring editor.

In the larger textbook firms, acquisitions editors specialize in particular disciplines, and may use the title of their discipline: biology editor, mathematics editor, economics editor, etc. In smaller houses, the acquisitions editor may actually be the president of the firm, or may have a general title such as publisher, managing editor, or senior editor.

Titles are deceiving. Within the company, the role of each person is generally known, but the author looking in from outside may have difficulty evaluating the acquisitions editor's power and status. External indications may help, such as the apparent age of the the editor, how long he or she has been with the company, the size and location of the editor's office, who answers the telephone (the editor or a secretary), the apparent size of the expense account, etc.

A key factor in evaluating status is whether the acquisitions editor is empowered to sign contracts. Most are not, and any promises made to an author will have to be cleared by management. Since the acquisitions editor is the author's main entree into the company, a powerful acquisitions editor is a major plus. Although the author can have a good relationship with a junior acquisitions editor (who may in the long run work harder to bring the book to publication than an overworked assistant vice president), there are important publishing matters which depend on the power which the editor has within the company.

Duties of the acquisitions editor

In a textbook house, the acquisitions editor's duties are reasonably well defined. They can be listed briefly:

- Keep informed of developments in the field
- Develop plans for books that are likely to be profitable
- Locate professionals in the field able to serve as advisors on manuscripts at all stages of development
- Make contact with potential authors, through travel to professional meetings and visits to college and university campuses.
- Work with field representatives to locate potential authors
- Induce authors to sign contracts
- Ensure that commissioned material is delivered by authors on time
- Receive completed manuscripts and determine when they are ready for editing
- Send manuscripts out for professional review, and evaluate the reviews
- Prepare a preliminary budget for each new book
- Determine when a book is ready for production
- Work closely with production editors in the preparation of the manuscript for production
- Consult with production editors and book designers regarding the costs and physical aspects of the book
- Work with sales and marketing departments regarding advertising and distribution of the book
- Be aware of activities of competing publishers
- Help in the preparation of sales literature for use by the field representatives
- Serve as liaison with the author, maintaining good working relations
- Read and evaluate unsolicited manuscripts
- Keep informed of the sales of books currently on the publisher's list and decide if and when revisions are needed

What are the backgrounds of acquisitions editors?

It will be rare that an acquisitions editor has any significant background in the discipline. This means that all of the technical terms that the academic author understands so well will be virtually a foreign language to the acquisitions editor. When the author carefully explains that the proposed text has major new material in "microeconomics" or "cell biology" or "relational data base management systems", the acquisitions editor will probably not look blank, but will certainly not understand in an entirely intelligent way the topic under discussion.

According to one survey of acquisitions editors in science, only 25% had a background in science, academia, university, or education. About 30% began as copy or manuscript editors, another 20% came out of the sales department, and a small percentage came from advertising, production, or professional fields. It seems to be well established in the publishing field that the acquisitions editor need not have any firm background or training in the academic discipline. In fact, an acquisitions editor will rarely remain in charge of one academic discipline for more than a few years before moving into another field. The following quotation is pertinent:

> The skills of the acquisitions editor . . . are interchangeable from discipline to discipline. I could just as easily walk down the hall and within two or three weeks pick up the physics editorship or electrical engineering and do almost as proficient a job as I do in psychology. From Coser, et al., page 103.

This philosophy is not held by all. Some acquisitions editors feel that specialized knowledge of the field is desirable. It is not so much the subject matter of the field that is useful, but the contacts with individuals in the field. An acquisitions editor builds up a list of names and connections, people that can be called on for advice about authors and books. An important part of the acquisitions editor's job is an evaluation of the market, knowing the course adoption potential of various books. This presumably requires some intimate knowledge of the field.

In my experience in dealing with acquisitions editors, I never knew one who had even an undergraduate degree in the field that I work in. Mostly, their backgrounds have been in humanities, business, or psychology, whereas I write books in biology. Nor have I ever had the same acquisitions editor for more than a few years. Mostly they move from one company to another, occasionally moving up within a company.

In some fields, technical expertise may be more common. In medical textbook publishing, most acquisitions editors have medical degrees, and with some technical publishers, engineering degrees are common. Acquisitions editors for some of the smaller science publishers have undergraduate degrees in a science, although not necessarily in the science for which they are acquiring titles.

Although there may be big money in publishing, acquisitions editors are not the ones making it. Their salaries are generally low, about what would be expected for lower to middle management in any large corporation. There may be some "perks", such as expense accounts, the chance to travel to interesting places, and the opportunity of working with intelligent people. But an acquisitions editor will not get rich remaining in that job. This is something the author should keep in mind when contract negotiations are in progress: the acquisitions editor may be offering the potential for more money than he or she will likely ever see personally.

Working with the acquisitions editor

I start with the assumption that a contract has been signed and the book is in progress. (Contract matters are discussed in Chapter 9.) The author is putting a major investment of time and nervous energy into the preparation of the text, and it is essential to be certain that the work is not wasted.

First, it is vital to remember that the publisher is under no obligation to publish the book, even if the author has complied with all of the clauses in the contract. If conditions have changed at the company or in the field since the contract was signed, the book may no longer fit the marketing plans or the mission of the company. Thus, it is vital that the author make frequent contact with the acquisitions editor during the preparation of the book manuscript, so that changes in the company that might affect the book are detected. Outlines, chapters, and letters should be sent at regular intervals. If responses to these items are long in coming, something upsetting may be afoot. If the acquisitions editor departs, as part of another publisher's "musical chairs", frequent correspondence will alert the author to this fact sooner, and the author will be able to develop contact with the editor's replacement sooner. It must be remembered that until the manuscript is delivered, the only concrete evidence the publisher has of the existence of the book is the signed contract, which is carefully filed away somewhere.

When attending professional meetings at which the publisher has an exhibit, the author should visit the booth and talk to the personnel. The acquisitions editor who arranged the contract should be in attendance. If not, why? If the author has been blessed with a large advance payment at the time the contract was signed, less trouble may be experienced in maintaining contact, because the acquisitions editor will have a financial interest in keeping track of the author's activities (the company already has something invested).

One point is that the acquisitions editor is involved in many books, and may lose sight of a book unless kept informed. If nothing else, at least once a quarter, the author should write a letter with information on the progress of the book (unless there is no progress; in that case, the author should lie low!).

Once the book is completed, the author should follow all of the publisher's instructions (see Chapter 5) and deliver the manuscript to the acquisitions editor. The author should call first and see if there are any special instructions about how it should be sent. If an agreed delivery date was specified in the contract, and there was a cash advance, the author should be certain that an acknowledgement is received in writing. I once had the experience of delivering a completed manuscript on schedule only to find that the acquisitions editor had suddenly left the company and had not been replaced yet. My manuscript gathered dust for months while a new person was sought and until things were sorted out.

Reviewing the manuscript

Once the manuscript has been delivered to the acquisitions editor, it will be logged in and steps will be taken to have it reviewed. No matter how well respected the author is in the field, no acquisitions editor will initiate the publishing process without having the manuscript reviewed by at least a few people. If the book is to be a major text with a large financial investment in production, the publisher will probably have it reviewed by a large number of people. It is also an excellent idea for the author to have the book reviewed independently of the reviews the publisher obtains, calling on knowledgeable colleagues.

Outside reviewers

Several types of reviews can be obtained, each serving a different purpose.

- Students. One or two senior or graduate students of the author's choice could be hired to read the manuscript and review the material with the author. The function of these reviews would be to read for understanding and to point out words, paragraphs, or concepts that are not clear.

- Specialists in the field. Active scholars would be hired to read specific sections dealing with their areas of competence, to assure the author that the most up-to-date and accurate material is included. These specialists would also offer a general evaluation of coverage and suggestions regarding organization.
- Series editors. Some publishers employ series editors, faculty members whose function is to help identify suitable authors and to advise the acquisitions editor. Such series editors might also be asked to review the overall manuscript for content, as well as to make detailed comments in their own areas of specialty.
- Potential users. Active teachers in schools of various types (including junior colleges, four-year colleges, and research universities) would be hired to review the material to assess the general effectiveness of the book as a teaching tool.
- Product management team. Some publishers have a special team employed to assess the sales appeal of selected books that are under development. The product management group makes specific calls on potential users and others knowledgeable in the particular market. The author is provided with information regarding the strengths and weakness of the book in relation to the competition. This information is most useful for final revisions and for identifying sales features that might be used by the advertising department and sales staff.

Reviews thus have three major purposes: 1) To determine that the book is worth publishing; 2) To obtain information on the likely market for the book; 3) To detect technical errors and errors of fact.

Although the third purpose is of most concern to the author, the first two are what the publisher most wants to learn from outside reviews.

For technical reviews, each chapter should be sent to an expert in the subject dealt with in that chapter. Rarely will the editor be able to determine the names of technical reviewers, and it is the author's responsibility to make these suggestions. Even if the author thinks that all the errors have been caught, it is inadvisable to avoid having the book reviewed by experts. There are simply too many possibilities for error, and when the book is subsequently reviewed in professional journals, major blunders will certainly be cited by book reviewers.

Reviews should be done at various stages during the development of the book, from initial conception to final manuscript. Once a detailed outline has been prepared (see Chapter 5), the editor should have it reviewed by individuals teaching appropriate courses. Although the author may be extremely knowledgeble in the field, it should be remembered that the author is not the one making the adoption decision. If the book is to sell well, then the desires of users must be taken into consideration. Once the outline has been refined and the author has begun to write in earnest, the editor should want to have some specimen chapters reviewed for content and style. The main point at this stage is a determination of whether the author can really write, and at what level. It is difficult to take a body of material and write it in such a way that it is intelligible for beginning students. We will discuss this in detail in Chapter 5. In a major textbook project, there should be reviews at all stages up to the final manuscript. For an untried author, much of the above may come before the contract is signed.

The budget that the acquisitions editor has for reviewing the book will provide clues as to how important the company thinks the book will be viewed. A chapter-by-chapter review, plus reviews of the whole book for marketing purposes, is an expensive proposition. Most publishers try to get reviews done inexpensively, paying professors $25 to $30 a chapter, or $300-500 per book, although higher payments may be made for major projects. Although such amounts are paltry when compared to normal industrial consulting payments, they are probably as much as most publishers can afford, since all reviewers' payments will have to be earned back by subsequent sales of the book being reviewed. Whether or not a useful review will be obtained will depend more upon the diligence and sense of responsibility of the reviewer (or perhaps on how well the reviewer knows the author personally), than on the amount of money changing hands. The author should feel fortunate when the editor has obtained knowledgeable reviewers who really tear the manuscript apart.

The reviewing process can take weeks to months, depending upon the agreement that the editor has with the reviewers, and how diligent the reviewers are. Since textbook publishing schedules are very critical, it is important that the reviewing process not hold up production. On a major project, the editor may be willing to pay more for quick turnaround on reviews, and the author should keep informed on how the reviewing process is to be scheduled. One desirable procedure for reducing reviewing time is to have chapters reviewed as they are completed. To do this, the author should have an arrangement ahead of time with the editor, so that as chapters or parts of the book are completed, they are sent out for review.

In addition to the reviews that the publisher makes, the author will probably want to have colleagues review particular sections for accuracy. For this purpose, the author would be sending the material to the colleague, and receiving the review back. Such reviews, not being anonymous, may be less frank than the publisher's reviews, but will probably be more useful from a technical viewpoint, since the author will know that the reviewer is an expert in the particular topic being reviewed. The author may sometimes find that colleagues do only a perfunctory job, but often they do extremely careful and very useful reviews. If prior arrangements are made with the publisher, these colleagues can generally be paid at normal reviewer's rates. Most colleagues will review material gratis as a favor to the author (but the author should ask the publisher to pay anyway).

When the publisher-requested reviews come in, the acquisitions editor will then be faced with the problem of evaluating them. The editor will be reading the reviews primarily to determine if the author is on target for the market, and if the book is likely to sell well. Technical reviews will not be understood by the editor, who will likely simply pass them on to the author. Such technical reviews may or may not influence the editor's decision to publish. Frequently, the editor will not be able to decide from the reviews whether the book is or is not worthy of publishing. It is at this time that the editor will call on outside advisors or series editors that the publisher has under contract. If such outside advisors are used, the decision to publish may well involve a lengthy discussion between the outside advisor and the acquisitions editor (and probably, if the acquisitions editor is fairly junior in the company, the editor's boss). I should emphasize again that the publisher is *not* under any obligation to publish the book, even if it is under contract and even if all of the requirements of the contract have been met. There is a clause in every contract, the "satisfactory manuscript" clause, that gives the publisher an escape route. If the manuscript is deemed to be unsatisfactory, the publisher can refuse to publish and all royalty advances must be returned. Although this clause has legitimate justification when used to avoid the publication of a disastrous book, it has also been used to get publishers out of expensive projects that may not suit new or changed corporate goals (see Chapter 3).

Handling bad reviews is a touchy matter. The editor, of course, hopes that all of the reviewers' comments will be taken into consideration and the manuscript will be revised accordingly. The author may feel otherwise. From the publisher's viewpoint, if all of the reviews are negative, some major overriding considerations will have to be present in order for the book to be put into production. Perhaps it is the only book in a new field in which a large market has just developed. Perhaps the author is so well known that name alone will sell a large number of copies. But generally these factors are not involved. The editor knows that the book will meet competition from similar books from other publishers. If a major financial investment is to be made, the book cannot be uniformly viewed in a negative manner by all reviewers. Hopefully, the author will be compliant, take the reviewers' comments into consideration, and revise the manuscript accordingly. The bad reviews should be put away for a week or two, then examined again to see if they can be accommodated. It is most important at this stage that the author not get discouraged by bad reviews. The author has put much time and energy into the project. To give up now would be to waste all that effort. If a microcomputer or word processor has been used (Chapter 8), at least the typing of the revised manuscript will not be too great an effort. Hopefully, sufficient reviews have been obtained at the outline and sample chapter stage so that the author has already tailored the writing to the market.

The decision to publish

Once all the reviews are in, and the acquisitions editor has had a chance to digest them and consult with outside advisors, inside colleagues, and supervisors, the decision to publish the book must be made. Before this decision can be made, the acquisitions editor, in consultation with the production manager, will have to prepare a preliminary budget for the book. A book budget is actually a complex document, since it must take into consideration all cost factors, from author's royalty through warehouse charges. We discuss these matters in Chapter 6. We note here, however, that if the book cannot be produced within a reasonable budget, it will not be a viable commercial prospect. Critical here is the length of the book and the complexity of the illustrative and tabular material. At this stage, some preliminary considerations about design of the book will also have to be made, because design considerations can greatly influence the budget. If two-color printing, or a large number of color plates, or a major investment in new art is suggested, then these design factors will have to be taken into consideration in developing the budget for the project.

Once the decision to publish is made, a whole new train of activities is set in motion. The production department takes over, and the acquisitions editor takes more of a background, advisory role. In some of the larger publishing firms, once the decision to publish is made, a "launch" meeting is held, at which the acquisitions editor meets with production and marketing people to discuss the nature of the book. At this meeting (which the author will not be invited to attend), the manuscript will be presented and the preliminary budget which the acquisitions editor has developed will be discussed from the viewpoint of production and sales. A production editor will be assigned to the book and a tentative time schedule for the book adopted. The author will know that the book is finally on its way when the production editor calls or writes.

Because companies vary considerably in how the publishing decision is handled, the author should ask the acquisitions editor how it is done. When and how is the decision to publish made, by whom, and using what criteria? It is also to the author's benefit to try to find out what the budget for the book is like. Critical here is a consideration of how many copies of the book will be made for the first printing, because the first printing is what the whole budget is based on. If the first printing is high (20,000 or over), then a nice, expensive budget can be developed, making possible the production of a very handsome book. If the first printing is low (10,000 or less), the production of the book will likely be run-of-the-mill. In most cases, the first printing represents what the publisher estimates will be sold in the first year the book is on the market.

The first-printing decision made at this time is only approximate, and will be refined as production proceeds and the production costs become clearer. However, most publishers have considerable experience with marketing their books, and the first-printing estimate will probably come fairly close to the mark. Hopefully, the book is one that justifies a large first printing, presaging a profitable book and a handsome author's royalty.

Editorial and production

Once the book has been put into production, a whole team of editorial workers takes over. The responsibility of the production department is to convert the raw manuscript into a form that can go to the printer. In a large publishing firm, the production department will employ many people, and will probably be highly organized. A production editor will be assigned to the book and will work closely with the author in guiding the book through the complex production process.

In some publishing companies, unusually large and expensive textbooks are handled by a special production team (which may be called something like "special projects" or "product development"). If a book has been chosen as a special project the author will know that the book is being viewed by the publisher as a "big book", with large anticipated sales. The production job done on such a book will almost certainly be of higher caliber or more sophisticated than if it were done by the regular production department. Special projects groups handle fewer books than the regular production department, work more as a team, and are probably more highly paid. A bad book can still result from a special projects approach, but the myriad details involved in book production are usually handled better.

Most books do not get the special projects treatment but are handled by the regular production department. Hopefully, the production editor assigned to the book will be good, will have been with the company for a while, and will not leave in the middle of the project for another position. The author will be working very closely with this individual over the many months required for production of the book, and although the two may perhaps never meet face to face, they will get to know each other well.

The main job of the production editor is the coordination of the other people involved in the production of the book. Although a number of people will be involved in one way or another, the main parties are: Copyeditor, who marks up the manuscript for the typesetter; Designer, who makes the decisions on size, shape, type specifications, paper, etc.; Art editor, who works with the artists and photo researchers; Manufacturing manager, who orders typesetting, printing, and binding. In a small company all of these tasks may be handled directly by the production editor, but in a large textbook firm, separate individuals are involved and the production editor's responsibility is to keep them moving forward.

We discuss textbook production in some detail in Chapter 6. Here, our main interest is in how the problem is organized at the publisher's level. One major matter of organization that often seems to surprise authors who have not published books before is how little of the actual work on a book is done by the publisher's own staff. Rather, it is done by outside suppliers and free-lance workers who are contracted for specific tasks. The publisher's staff works mainly to coordinate the outsiders, and to decide who to hire. Although the author may be sending manuscript, galleys, art, page proofs, etc. to the publisher, the publisher is just a transfer point between the author and outside suppliers.

Marketing, sales, and advertising

As the book moves through the later stages of the production cycle, the marketing department will begin to get active (see Chapter 12). It is the marketing department's job to develop a marketing plan for the book. Marketing is one of those professions that is difficult to define, even though it pervades modern business. It is often difficult to separate marketing from sales on the one hand and advertising on the other. To a first approximation, marketing can be viewed as the coordination of sales and advertising to bring a product to market. Although consumer-product firms spend large amounts of money on sophisticated market research to determine what the public will buy, publishers rarely do anything more than rudimentary market research. Also, the marketing of textbooks differs considerably from the marketing of trade books. With trade books, there may be considerable author involvement in marketing (including promotional events such as author's tours, appearances on television programs, autograph parties, etc.), whereas with textbooks, the author will be involved only peripherally.

The title of the marketing department varies from publisher to publisher. The department may have the word "promotion", "publicity", or "marketing" in its title. In one marketing department, the duties include: plan the advertising and promotion programs for each book via consultation with the editorial and sales departments; prepare the copy for each textbook listed in the publisher's catalog; edit all copy provided by the advertising and editorial departments; work with printers on production of direct mail and catalog materials; work with sales and data-processing departments in the acquisition and development of lists for direct-mail advertising; coordinate the actual mailing of direct-mail pieces; work with the sales department on the actual sales of each book. In addition, the marketing department will work with the acquisitions editor in the early stages of developing a specific book project, even before the contract is signed, to be certain that the proposal is economically viable.

Advertising may be prepared by the publisher's own advertising department or by an outside advertising agency. With textbooks, outside advertising agencies are not often used because the requirements of textbook advertising are too specialized for the advertising agencies. Advertising includes both direct mail and space ads in journals. Depending on the budget for a book, there may be a direct mail advertising piece specifically for that book, or it may be advertised together with a group of related (and noncompeting) books. Space ads in journals will almost always involve a group of related books, unless the book is considered to be a real blockbuster.

The author may or may not be specifically asked to read advertising copy. Because advertising copy writers often have little appreciation for the nuances of academia, some of the copy they write is really awful. The author should insist, and have it written into the contract (see Chapter 9), that all advertising copy *must* be approved by the author. The more the author provides the advertising department in the way of information or potential copy, the more likely the final advertisement will be satisfactory (see Chapter 12). The acquisitions editor should be informed that the author is to be involved in all advertising copy writing, and as soon as the first communication from the marketing department is received, this department should also be informed. Copy for the publisher's catalog is not as important as direct mail and space advertising, since this is not seen by the author's colleagues but only by bookstore and sales personnel. Nothing can be more embarrassing than a poor advertisement sent to all the members of the author's professional society. Don't let it happen!

For the larger textbook publishers, the sales department employs many field representatives (sometimes called "college travelers", "detail men", or simply "reps"). A professor who teaches a large-enrollment course has already met many of these people. It is their responsibility to get the book into the hands of the proper professors, and to attempt to induce them to adopt it for their course. The value of field representatives for the sale of a textbook is frequently debated. The large textbook publishing houses have many field reps, and they certainly sell lots of books. Most people in textbook publishing agree that the single most important factor in selling a text is getting it into the *proper* professor's hands. Because of the complexities of academia, the responsibility for the major courses changes hands frequently. The field representative, being on the spot on the college campus, may be able to ascertain who is teaching the course this year (or, even more important, next year), and get the new text into this person's hands.

However, the field representative has many new textbooks to sell at any given time. Whether or not the field representative will push the book will depend partly on whim, partly on luck, and partly on how well the book has been "hyped" by the acquisitions editor and sales manager. A big thing in publishing circles is the national sales conference, where sales personnel and editors from a given publishing house get together to discuss the forthcoming books. The acquisitions editor (who may well have begun as a field representative) will attend the national sales conference and will be pushing books that he or she has been responsible for bringing out this year. If an author is lucky, the book will get a big play at the national sales conference, and the field representatives will get enthusiastic about it, and push it. Since the field representative is paid partly on the basis of how many adoptions he or she obtains, there is a strong interest in pushing books that are likely to sell well.

An important thing that the author can do is get to know the local field representative. If this person gets to know an author and gets enthusiastic about that author's book, some of this enthusiasm will be communicated to the reps colleagues at the sales conferences, and this may lead to better sales.

Another role of the field representative is working with the college bookstore to be certain that the book is available when needed. Who has not had the experience of adopting a new book, only to find that it will not arrive until the middle of the semester? Although delivering books to bookstores is the responsibility of the order fulfillment department (generally located in a warehouse somewhere in New Jersey), the field representative can often expedite order shipping. I have heard of cases where an adoption of 500 books would be assured provided the publisher could have the books in the bookstore's hands within two weeks. Shipping that many books that quickly is a major shipping problem, but a good field representative can help in such a matter.

Managed texts

A phenomenon of recent years in the textbook publishing field is the managed text, a book without an author. A managed text is one written and produced by a team under very close editorial control of the publisher. The academic people involved in a managed text are hired not for their writing or organizational ability, but for their knowledge in a particular part of a field. In most managed texts, the final writing is done by professional writers (some of whom may be failed academics), using detailed notes or preliminary text provided by academics. If there is an author's name on the title page, this author may have functioned primarily as an academic consultant, a critical reviewer of all of the material provided by the writing team. Writing a textbook from material supplied by others must be similar to writing a travel guide to a country without ever having visited the country, using notes provided by others. It can certainly be done (many travel guides are actually written this way), but what happens to the flavor of the field, the nuances of expression?

An important distinction between a managed text and a conventional textbook is that the managed text is considered a "work for hire" in the legal sense of the copyright act (Chapter 10). The writers and academics involved in the production of the managed text are usually paid flat fees rather than royalties, and the copyright is obtained in the name of the publisher.

The advantage to the publisher of the managed text is that the product is under close editorial control. If the market research that the publisher carries out indicates that the textbook for a given course should have a particular set of elements in order to be adopted, then it is possible to ensure that these elements are included in the book. No recalcitrant author needs to be wheedled into providing the needed material. The large investment needed to bring out a textbook for a major course can be properly directed at the necessary elements of publishing success.

For the professor participating in a managed text process, the advantage is that little intellectual or emotional effort is involved. The professor is usually being asked to provide material for a segment of the book that relates directly to his or her specialty, so that the material can be supplied almost without outside reading. The disadvantage is that there is very little money or prestige involved, and the professor may only be listed in the textbook as one of many "advisors". In fact, many of the academic participants in a managed text may be graduate students or nonfaculty professionals such as postdoctoral fellows, hired for a few dollars.

The managed text often has a tell-tale appearance about it. The production job is generally superb, with four-color illustrations throughout, flashy and colorful diagrams, special boxes of "relevant" material, and a strong, handsome cover. A lot of money has obviously gone into the book, and it shows. Whether the book "works", in the educational sense, is another matter. Certainly a managed text that hits the market at the right time with the right material will sell copies. But a key factor in textbook publishing is continuity from one edition to the next. If a textbook is successful in its first edition, it will be easy to sell second and subsequent editions. With managed texts, being handled primarily by free-lance workers, this successful continuity may be difficult to maintain. It is also frequently the case that the managed text, although well-written from a professional writer's point of view, does not sit so well with the student.

There was an important legal case of a managed text involving plagiarism, where one of the "professional" writers copied material directly from the textbook of a major competitor, leading to a final settlement to the publisher of the plagiarized book of $487,000. The publisher, not the author, had to pay, since the managed text is a "work for hire" in the legal sense, making the publisher, not the author, liable.

The author interested in writing any textbook should have at least a little familiarity with managed textbooks, if for no other reason than to be aware of what the competition is all about. An excellent discussion of managed texts can be found in the book by Coser et al. cited in the bibliography.

Book packagers

A book packager is an individual or organization that puts together a book for a publisher, delivering the book to the publisher in final form, either as camera-ready copy or as final books. The packager finds the author, handles all editorial and design matters, and contracts with artists, typesetters, and others for the production of the book. The role of the publisher is primarily marketing and sales. Book packagers are very common in trade book publishing, less so in textbooks. The managed texts discussed in the previous section may be done by book packagers instead of by the publisher's own staff.

The advantage to the publisher of working with a book packager is that all of the difficult work is done. Also, all of the financial investment in editorial and production matters is made by the packager, so that the publisher has much less money involved. Of course, the packager must also receive payment, and depending upon the financial arrangements (and there are many different possibilities here) this may come out of the author's royalties. What are the advantages to the author of working with a book packager? The book packager may be much more knowledgeable in the particular area than the publisher's editorial staff, and may thus develop a better and more marketable product. If this is so, the higher sales may compensate for the lower royalty that the author may be receiving.

Any textbook author entering into an agreement with a book packager should find out in detail all of the financial arrangements involved before signing a contract. Who is paying for what? Does the packager already have a signed contract with a publisher, and what are its terms? Who will hold the copyright on the book? A packager may not want to provide some of the financial information, but without it, the author is entering uncharted waters.

Final words

In this chapter, we have presented briefly the various steps in the publishing process, and have concentrated on the acquisitions process. From the publisher's viewpoint, acquisitions is the key to successful textbook publishing, and the acquisitions editor is the author's central contact. The acquisitions editor is rarely trained in the discipline, but makes use of outside advisors and reviewers to evaluate the worth of a manuscript. The author should work closely with the acquisitions editor during the development of the project, to be certain that the final manuscript meets the publisher's requirements.

One point that deserves emphasis again: Despite the fact that a contract has been signed, the publisher is not obligated to publish the book if the manuscript does not meet current needs or if it is deemed unsatisfactory. Thus, to be certain that all of the work is not wasted, the author should keep close contact with the acquisitions editor during the development and writing of the book.

The publishing company is putting out many books. The acquisitions editor will be responsible for many books, and the editorial and production people will be working on more than one book at a time. It is very likely that no one in the publisher's organization will actually read any book that is being considered for publication. Indeed, except under rare conditions, the acquisitions and production staff will lack the training to understand the book. The only person who has likely read the book from cover to cover is the copy editor (see Chapter 6), who is marking the book for the typesetter and is reading it for grammar and consistency, but not for content. The outside reviewers do read the book, but frequently a single reviewer reads only a few chapters.

From the publisher's viewpoint, when the book is published it becomes a property or a product. Indeed, it is a book, but the publisher treats it as a commodity. It appears on computer printouts of the publisher's sales records, is discussed at sales meetings, and is shipped around the country in cardboard containers. This is the inevitable fate of any textbook good enough to get published.

From the author's viewpoint, the book is anything but a commodity. It represents a huge intellectual and emotional effort, and although the royalty checks may be welcome, that was not the main motivation for writing the book. What does make it all worthwhile is the sight of students carrying the book as they go to class, or to learn, years later, that a student's decision to enter the field resulted from the stimulation received from reading this textbook.

5
Writing the book

A contract has been signed with a reputable publisher and work is ready to begin. The author has been teaching the course for a few years, and has a good idea of what should be covered in the textbook. A preliminary outline was probably developed before the contract was signed. It is now time to get down to work.

One of the first things that must be understood is that the course that the author has been teaching cannot, by itself, be the basis of a textbook. Although in a given field most introductory courses have many common elements, each course is unique, adapted to local conditions. The author is not writing a book to be sold to the few hundred students taking the course at a single institution, but for the thousands across the country. The first job is to generalize the field.

The author must also remember that the book will be read by students but will be adopted by professors. Very few of the professors adopting the book will have read it before they adopt it. Rarely will there be a professor using the book who has read it from cover to cover. The book will be judged by the professor on the basis of impressions gained by turning the pages, looking at the pictures, and reading snatches of text here and there. Each professor will have prejudices about the course: what material should be covered, the organization of the material, the level of sophistication of the students. The author must ensure that all the elements of the course have been covered, that nothing significant has been left out.

Although adoption of the book may come without anyone having read it, the students *will* read it, and if they have difficulties, they will complain to the professors. The goal must be not just to have the book adopted, but to have it be used continuously through several editions over many years. The author must build up good will and a strong image in the professor's mind: this will be determined primarily by how the book is perceived by students. The author tries to strike a balance between the needs of the students and the attitudes of the professor.

Ignore competing textbooks

One of the first rules of successful textbook publishing is to ignore other people's textbooks. Once a decision has been made to write a textbook, other textbooks should be put away and not looked at again (except for the index, see below). The main reason for this is that the new book must look fresh and novel. Although the author may not consciously copy other books, there will be a strong urge to use illustrations, tables, or topics used in other books. The author may even find that the publisher is asking for the inclusion of certain things because they are present in a competing book. Certainly the author is not going to leave anything important out, but is also not going to put it in just because it is in a major competitor. If the author has been teaching the course, some familiarity with the leading textbooks already exists and this familiarity is all that is needed to get the new project underway.

It is most important of all to *never* use illustrations that have already been used in other textbooks, unless there is simply no other way to get across a point. In developing the book, the author must decide how important points should be illustrated. It is tempting to use or adapt a successful illustration from another textbook. Don't do it!

Developing the outline

One of the most important tasks in writing the book is developing a detailed outline. Editors will generally insist on seeing an outline, probably even before the contract is signed, and writing will be much easier if an outline is available to work from. Textbook writing is 50%

organization, 40% hard work, and only 10% writing ability. The outline is the key to good organization.

The author should begin by having a clear idea of the level at which the textbook is to be used. Many topics that might well be covered because they are interesting may not be suitable for a textbook because they are simply too difficult to get across to beginning students. The backgrounds of the students must be kept clearly in mind. If a biology textbook is being written, will the students have already had chemistry? If the author is doing an economics book, how much statistics and mathematics will the students have had? If the book is in sociology, can it be assumed that the students know some psychology?

What kind of course will the book be used for? In many universities, there are two levels of course, often with similar titles. One course is for majors and others with a professional interest in the field, the other for nonmajors, those who are taking the course as an elective. Certainly, a textbook for majors can be written on a much different level than a textbook for nonmajors, and can include more advanced material, even if the majors have never had a previous course in the field. As the outline is developed, the audience for the course must be kept firmly in mind. This will probably have already been agreed upon with the publisher, and should even be spelled out in the contract.

What length is the book to be? If a contract has been signed, this figure should already be specified. The outline for a short book must be much different than the outline for a long book.

Outlines of other courses

A very useful thing to do early in the development of the book outline is to obtain the outlines for similar courses taught elsewhere. Colleagues at other universities teaching similar courses should be quite willing to send their outlines. The publisher's marketing and sales staff (see later) will also be able to obtain outlines. These will give the order of topics covered, and some idea of how each topic is broken down. After 8 or 10 outlines have been obtained, patterns will emerge, and the key topics that everyone thinks are important can be highlighted. When examining these outlines, the level of the students in each course must be understood.

Editorial, marketing and sales input

If a contract has been signed and the outline for a major textbook project is being developed, the author should be receiving feedback from the publisher's staff at the outline stage. Early interaction between the marketing department and the author should be encouraged. As noted in Chapter 4, the publisher is not obligated to publish the book, once completed, even if a contract has been signed. To avoid a lot of wasted work, it is essential that the marketing and sales departments provide feedback to the author as the project progresses. They should be taking the outline into the field and determining how well it is perceived by potential users.

The ideal situation is to be working closely with an acquisitions editor who has taken a special interest in the book and who gives detailed criticism of the outline. If the book is to be a major textbook by a large publisher, and large amounts of money are going to be invested, a special projects editor may well be assigned to the book at an early stage to ensure that the whole project is developed properly. A small amount of money spent now on intelligent editorial advice can avoid wasting a lot of money later on the production of a book that does not sell. Publishers vary in how much they will invest in early editorial work, and this will also vary with the book. If an author is not getting much feedback, the acquisitions editor should be contacted and a procedure developed for obtaining feedback. A large advance on royalties is usually an assurance that the acquisitions editor will spend some time overseeing the project and giving the author feedback.

The index approach to outline development

Although I strongly advised against looking at competing textbooks, this does not include their indexes. In developing the outline, a very useful way of proceeding is to use the indexes of three or four competing textbooks to determine the range of topics covered.

The procedure is simple. The indexes of the key textbooks in the field are examined and the major and minor index entries are written on small slips of paper or index cards. If a secretary is available, the author can check the entries and the secretary can type them onto slips (or even into a microcomputer if appropriate software is available). If there are significant topics that are absent from all of the books, these topics should also be written on slips.

Once the items are all on slips, the slips from the various textbooks are collated and duplicates are discarded. At this point, all the elements needed to develop the outline are available. Because every item is available on a slip of paper, there is no danger that an important topic will be left out. The act of going through all of the slips will force the author to think about how different topics can best be fitted together.

A preliminary chapter outline may have already been developed before this exercise was begun. Or one may have been developed in a preliminary way as the indexes were examined. With this chapter outline written out, each slip can be located in its proper chapter. Some topics could be located in several places and the author will have to make some judgements. As the process continues, the slips can be moved around as needed. It is important not to rush through this exercise, because it is one of the most important things that can be done.

How long should the outline be?

Since the outline is going to be used by the editorial and marketing staff to get feedback from professors, it should be quite detailed. Every section head and subhead that would be included in the book should appear in the outline. It is also desirable if a word or two of explanation is provided under topics that are not understood immediately from the title. For a major textbook, any useful outline should be at least 15 pages long, preferably longer. It would also be valuable to include indications of what the illustration program will be like, the approximate number of figures or tables to be used in each chapter, and some examples of the kinds of illustrations to be used.

The outline should be preceded by an explanatory preface. The point of the preface is to explain what the book is trying to do, who the students using the book will be, their backgrounds, the level of the course. Prerequisites needed for an intelligent understanding of the book should be clearly spelled out in the preface.

Organizing the outline

College textbooks often have a sameness of organization that is conditioned to a great extent by the orderliness of the academic mind. It is nice to try to break out of this monotony, but it is also essential to have an organization that will be easy for students and professors to follow. There are ways in which a book with a standard organization can be made to look fresh. This is the function of the book designer, as we will describe in the next chapter.

At least the following elements are found in each chapter of a college textbook:

- Chapter number and title
- Introduction
- Major section head (called #1 head or A head by book designers)
- Subsection head (called #2 head or B head)
- Subsubsection head (called #3 head or C head)
- Summary
- Selected readings

Of course, for each chapter there will be a number of major section heads, and for each major section head there may be several minor section heads.

In some textbooks, additional elements will be found in each chapter. Some textbooks have an opening section called (in the best educational jargon) "behavioral objectives", which tells the student what he or she will (or should) know after having mastered the chapter. In many fields, there will also be lists of questions at the end of each chapter. These questions may contain problems involving calculations, multiple-choice questions of the type found on exams, or thought questions which are intended to get the student to think about the material in the chapter. Although these additional elements often help the sales of a textbook, they are often added last, frequently as an afterthought, or at the publisher's insistence.

When organizing the outline, it is very useful to number the major section heads. Thus, a major section head in Chapter 9 might have the number 9.4 or 9-4. Although use of section numbers makes the book look rather formal, they are of great use for cross-reference and avoid the vexing problem of cross-referencing by page number (the page numbers are never known until the book is finally produced, whereas the section head numbers are defined by the author as the book is written). Even if section head numbers are not used in the final book, they are useful in the outline.

Although numbers for #1 heads are useful, numbering of minor section heads is questionable. Systems of numbering such as 9.4.3 or 9.4B are sometimes used, and I have one book on my shelf which has subsubsection numbers, such as 9.4.3.2. To me, such formality is ridiculous, even in a mathematics textbook, and I would strongly discourage it. Even without the numbers, it may be questioned whether subheads less than the #2 are warranted. The book designer can specify different typefaces for the various levels of subhead, and this may help the people producing the book to keep things straight, but readers will have a hard time following all this jumble. My personal feeling is that if the material is so complex that subsubheads are needed, then perhaps the chapter should be broken down into two or more separate chapters, so that nothing less than a #2 head is needed.

Sources

If other people's textbooks are not being used, what are the sources of information? If the course has been taught for some time, the author probably has a fairly extensive bibliography built up of monographs, review papers, and specific research papers that deal with the subject. For the textbook, this bibliography will have to be expanded, because sections will be written that are not covered in the author's course, and more detail will have to be presented in many areas outside the author's specialty. The secret to writing a good textbook is to have a good grasp of the literature in the field.

The most useful sources for textbook writing are review articles and monographs dealing with specific areas. Journal papers based on original research are less useful because they are so specific, although they often provide good sources of illustrative material. The author should begin to collect for a personal library various monographs and other books (but *not* textbooks) that deal with the subject. These books will not necessarily be read cover to cover, but will be examined for specific points as the writing progresses.

The libraries in the local institution will of course provide access to virtually all of the literature that will be needed in writing the book, but it is very useful and an efficient use of time to have the key sources available in a personal library. The size of this personal library need not be large, but should be precisely focussed.

Some tools that are very useful, no matter what the field of study, are a good encylopedia, an unabridged dictionary, and professional handbooks. Since I write textbooks in science, I find very useful a 15 volume encylopedia of science and technology, handbooks of physics and chemistry, mathematical tables, and specialized technical dictionaries. My investment in these items has been paid for many times during my years of writing.

If the literature in the field is large, or is very broad, the size of this personal library may represent a significant financial investment. All of the expenses are deductible from the income tax. Also, if the author has obtained a good-sized manuscript-preparation fee from the publisher (see Chapter 9), the purchase of these materials is a legitimate use of that money.

The author probably has already developed a filing system for scholarly materials. It may now be useful to reclassify many materials into categories that relate to the chapter outline. Some sources may apply to only a single chapter, others to multiple chapters. A chapter-by-chapter filing system is often useful and makes it less likely that an important point will be forgotten during the writing of that chapter.

The actual writing

After all the preliminaries, it may seem almost anticlimatic to finally begin to write the book. If the outline is well developed, and if adequate sources are available, the actual writing should proceed relatively smoothly. The key problems involved in the actual writing are: finding the time, finding peace and quiet, getting the material down on paper, getting the material typed.

How long does it take to write a textbook?

Books differ, authors differ, and fields differ, but a new book begun from scratch should take between one and two years to write, if done along with one's other professional duties. How many hours of actual writing? An approximate breakdown was given in Chapter 2. I estimate about 2 hours a day, five to six days a week and about 100 weeks as an upper limit of required time. One person may prefer to write the book full time for a shorter period of time (I don't recommend it, but it is one way), and this may shorten the required time somewhat, but I still feel that 1000 hours is the minimum needed to accomplish the actual writing of the first draft of a major new textbook. (If the royalties over the whole edition are $100,000, then this amounts to a rate of $100 an hour, a reasonable rate of return.)

Scheduling the writing

Certainly, everyone must arrange their schedules to fit their own personalities and life styles. My preference is to write in short stretches, between 2 and 3 hours at a time, and do this regularly and consistently over a long period of time. I have written most of my textbooks between 5:00 and 7:00 A.M., getting up regularly *every* morning and writing diligently. To avoid any potential concern about conflict of interest with my university duties (see Chapter 13), I have done all the writing in my home office. (The home office may be deductible from the income tax.) It may not be someone else's idea of fun, but it results in finished books.

Where to write?

I have written in many places, both near and far away from home. Although exotic places sound attractive, I have always done my best writing in my home office. There are many reasons for this. Distractions are virtually zero. I have good supplies of paper, pencils, paper clips, etc. I have all of my sources ready at hand. The copy machine is nearby. I have my microcomputer and typewriter. If I need a secretary, I can easily hire one on a part-time basis.

Although it is certainly possible to write anywhere, I believe it is important from a psychological viewpoint to write close to my place of work. Reality is nearby. Daily contacts with students keep me informed about who I am really writing for. If I have a question, I can easily get it answered, rather than having to file it away for future reference. If I were working in a remote location, I would have to carry with me a mountain of papers, books, equipment, supplies. I would certainly forget something and would find myself compromising the project.

Getting the manuscript on paper

There is many a change between the outline and the manuscript. Things that looked good in outline turn out to be disastrous when committed to paper. Things that seemed important at the outline stage may turn out to be minor points that hardly warrant separate sections. It is important to understand that the act of writing itself calls forth new ideas and new approaches. One must be prepared to modify the outline, drastically if necessary, as the writing proceeds. It is also the case that the later chapters will be handled differently than the earlier chapters, because experience has been gained writing the earlier chapters and because the later chapters build on the earlier ones. The earlier chapters will thus have to be more heavily revised than the later ones.

If a microcomputer is being used (see Chapter 8), the author will be thinking about typing the whole book directly into the word processor. Perhaps this will work for some. It sometimes does for me, but more often does not. Although I can type fairly well, I find that writing complex material at the screen of a word processor is difficult. What I must do is first commit a draft to paper in handwritten form. I do this on large pads of lined yellow paper, keeping the sheets together on the pad as I write. This draft is extremely rough, and is being modified greatly as I go along. Then, when I get it in a form that I think I like, I type it on the word processor. I find that as I enter the text into the word processor, I rewrite significantly and refine the words and style. The handwritten version has served to get the organization right, and the order of topics right, and now I can get the syntax and style right as I put it in typewritten form.

Once the whole chapter or major section has been written on the word processor, I edit the printed version. Often, a significant amount of editing and rewriting is needed at this stage, but it is easy to do because of the convenience of the computer text editor. The version that has been edited on the word processor is then the one that will be go to reviewers.

Another approach I have used is to take my rough version and dictate it, modifying the text as I go. Then a secretary transcribes the material into a word processor and provides me with a typed version that I then edit drastically. Dication is markedly quicker than typing, even with a word processor, and is probably cost-effective, if my time is figured into the cost analysis. However, I find that I rewrite less when I dictate than when I type.

When the reviewer's comments come back, one must come to grips with what the final version will really be. If the author has made certain that good, highly critical reviewers have been obtained, as I advised in Chapter 4, then there will be many potential changes to make. Reviewers may suggest deletion of whole sections, or adding new sections, or restructuring of the chapter. Since the responsibility for the final book rests with the author, the author's judgement must rule in all matters of this sort. If the author has some perfectly valid reasons for doing things a certain way, or thinks that the reviewer has missed the point, then a reviewer's comments might be ignored. However, the author should not forget that if the reviewer has some difficulty with the material, then students may well have difficulty also. If a reviewer has flagged something in the manuscript, it is very likely that there is a problem there that needs attention. The reviewer's suggestion may not actually be the correct one, and it is the author's reponsibility to decide what to do. Certainly, if several reviewers comment on the same point, that is a good indication that the author has really gone astray and should rewrite.

If the author has a fragile ego, it is most likely to get bruised while reading reviewers' comments. The author should not get discouraged. The reviews should be scanned first to see how bad they really are, then be put away for a day or two. When beginning to go through the reviews in earnest, the author will have already developed a protective shell and can accommodate to the reviewers' comments in a reasoned way. At least the author has the knowledge that this is the final revision of the book, and from here on it is downhill to the printer.

The textbook style of writing

Most of your professional career has involved writing and reading monographs and research papers. You read very few textbooks, usually not even the ones you use in your own teaching. One of the most difficult tasks is to shift into the textbook style of writing. The success of your book may well depend on how well you do this.

One of the first things to remember is that the students reading your book are learning about the subject for the first time. They have no previous background in the field. They are not aware of old ideas and misconceptions in your field that have been discredited. You do not need to list every theory or explanation for a given topic, but only the currently accepted one. Your task is to present the field as it is today, the "state of the art". If your book is successful, there will be more editions, and current dogma can be replaced as it is discounted or discredited.

The textbook style of writing is direct, didactic, well focussed. Here are two contrasting styles of writing, one textbook, the other not:

> We have shown that the uplifting of land masses has led to the formation of mountains. We now discuss the volcano, another type of mountain, formed in a quite different way. A **volcano** is a mountain formed when magma or rock from within the earth is forced to the surface and piles up. There are several kinds of volcanoes, but all have their origin in this extrusion of material from within the earth. We list briefly the main types of volcanoes, and will discuss the manner in which each type is formed.

> The visitor arriving in Hawaii by airplane lands across Pearl Harbor from an ancient land mass that was formed millions of years ago. The rock is reddish and jumbled, and the vegetation hangs precariously to the steep cliffs. Climbing to the top of this craggy peak, the visitor is startled to find a deep pit which looks as if it descends right into the bowels of the earth.

> The ancient Polynesians actually believed that this mountain was the entrance into the lair of Pele, the goddess of the earth. This is a fascinating example of another type of mountain, the volcano.

Clearly, the first paragraph is textbook style, and the second is not. Although the second paragraph may be more fun to read (I make no strong defense of the writing style), it tells the student practically nothing about volcanoes.

The textbook-style paragraph has several key elements that should be emphasized. Its first sentence relates back to previous material that the student has already mastered, thus providing a point of reference. It is already clear from this sentence that we are continuing the topic of "mountains", so that the student knows precisely where he or she is. The second sentence clearly states that we are beginning a new topic, and tells us what that topic is. A new term is introduced and *immediately* defined in the following sentence. The word being defined is printed in **bold** type, and the whole sentence stands out clearly because of this. When scanning back over the material later, the student can easily find the sentence in which the word was defined. The next sentence leads forward some more, telling the student what is coming next and getting ready for further new words and ideas.

The style of writing is perhaps dull, and may not be very much fun to write, but it is the style needed if the book is to serve the student well. The narrative style of the second paragraph can sometimes be used successfully under the proper conditions. Generally, narrative is more suitable for material included as "enrichment", perhaps in a separate box, often as a trailer paragraph to a section in which the key terms have already been well explained.

A third style of writing frequently used in textbooks is the "historical" style, which attempts to show the student steps by which currently accepted wisdom was developed. Here is an example of historical style:

> What are volcanoes composed of? At first it was believed that volcanoes were composed solely of basalts, but now we know that volcanoes can be composed of either andesites, basalts, or tufa. Thor Jonsson, the famous Icelandic volcanologist of the late 19th century, first described tufa as a volcanic rock. Later workers, visiting only Italian volcanoes, disagreed with Jonsson's conclusion and a controversy developed which raged until the 1960's. In 1963, the famous eruption of Surtsey, off the south coast of Iceland, permitted a direct study of the formation of volcanic tufa, and all workers now agree that tufa is a volcanic rock.

This paragraph is an example of the historical approach that leads to a verbose and lengthy book, but which may leave the student little enlightened. Such a style might not even be appropriate in a history textbook!

It should be remembered that the book is written for students who know nothing about the subject. The student should always be aware, when reading a paragraph, precisely why that particular paragraph is being read, why it is being included in the book, what its purpose is. The textbook is not a mystery story. A direct, forceful, uncomplicated approach is needed.

However, this does not mean that the writing must be dull and formal. There are plenty of opportunities for enlivening the material.

Although it is hard to accomplish, the author should strive to make the book interesting for the better student, yet accessible to the poorer student. This can best be done by keeping a highly ordered text, with clearly delineated topics and precise definitions of terms, while adding the more interesting material as supplementary topics after the key points have been covered. The good student will breeze through the simple material and will then stop to enjoy the enrichment material. The poorer student will struggle with the main subject matter and will merely skim or skip entirely the enrichment material. The poorer student will be studying the text primarily to learn the material needed for the examinations, whereas the good student will be studying the text primarily because of an inherent interest in the material.

Some helpful hints for textbook writing

Transitions are very important. The student should always know precisely where he or she is in the book, and why any particular paragraph is present. Here are some useful transitional phrases:

We thus see
We now consider
In the last section we
We list briefly
However, this does not mean
Likewise

Ask a question and then answer it. The student is pulled along by the question, pauses briefly to consider, then consumes the answer.

What is a volcano? A volcano is . . .

Make judicious use of lists. Two kinds of lists are used, numbered and unnumbered. An item in an unnumbered list is usually introduced with an ornament, such as a bullet • or a check ✓. Lists are useful because they permit the student to see in one place a group of related items. However, the items in lists should be short, no more than a sentence or two. Although lists are useful, one danger of lists is that students often feel that they have to memorize them, or that there is something unique about them.

Break the text into relatively short elements. Nothing is worse for the student than a long paragraph that seems never to stop. Short paragraphs, section heads, the frequent use of figures and tables, typographic devices such as boldface, all help to keep the student awake and interested and to keep the book from looking dull.

The author should be certain to keep all promises. If the text says that something is going to be covered later, be sure that it is covered.

It is inadvisable to use the first person. The textbook is an impersonal object and the student will care little about the author's personal ideas. The second person should also only be used sparingly, if at all. A perfectly good direct style of writing can easily be accommodated by writing in the third person.

Grammar and related matters

A large number of problems come up in writing a textbook that deal with grammar, syntax, spelling, punctuation, proper abbreviation, hyphenation, etc. Although the textbook should be grammatically correct, it is not critical at the initial writing stage to be too concerned with these matters. Most author's guides cover grammar in great detail, and the number of rules and standard practices are so numerous that a conscientious author might never get to the actual writing for fear of making a mistake. It should be remembered that the manuscript will be edited by a copy editor, whose job it is to worry about these things. Many publishers have their own style sheets or manuals that discuss most of the important grammatical problems, and it is the copy editor's responsibility to make the manuscript conform to this style.

For occasional reference, one of the many style manuals available at the college bookstore can be used. I list several in the bibliography, but note here that a style manual frequently used by copy editors is the *Chicago Manual of Style* published by the University of Chicago Press. This book is especially good for scholarly style in humanities, social sciences, mathematics, and psychology. It is less good in business, engineering, and the physical and biological sciences, for which the McGraw-Hill style manual may be preferable.

I have found through the years that I rarely use any of these style manuals. What I do use frequently is the unabridged dictionary. I need this not only for checking the spelling of unfamiliar words, but for determining whether a word I think I want to use is really the proper one. Although there are special dictionaries in the fields that I deal with, I have found that the unabridged dictionary covers virtually everything the specialized dictionaries do, but also includes the non-specialized words that I frequently need to look up.

Some general rules for textbook writing

Although textbook writing is similar to writing any other material, and the basic rules of good writing apply, there are certain rules that are more specifically applicable to the textbook.

- Technical words should always be defined the *first* time they are used.
- A complicated word should not be used when a simple word can be used instead. Why use "neologism" when "new word" means the same thing and is more understandable?
- New words should not be coined. A textbook should reflect the state-of-the-art, but should not be ahead of it.
- Trite, slang, or loose expressions should be avoided. While one person may use the word "acid test" to refer to the measurement of the acidity of a solution, another may use it to refer to a critical trial. The latter usage is not desirable in a textbook.
- Nouns and adjectives should not be used as verbs, unless the usage is well established. Although "formalize" may be acceptable, "rigidize" is definitely not ("make rigid" does it so much better).
- Nouns should not be used as adjectives unless the usage is well established. This fault is especially serious if several nouns are strung together. For instance, the phrase "heart chamber pressure change" would be more readily understandable if it were expressed as "change in pressure of the chamber of the heart". If the longer phrase seems to awkward, then a judicious use of hyphens permits "heart-chamber pressure change", which is clearer.
- Pronouns whose modifiers are unclear present one of the worst problems in textbook writing. Especially difficult are "they", "that", "those", "this", and "these". Consider the sentences: "Although the use of section heads and part titles makes the book look rather formal, *they* are of great use in . . . " What does "they" refer to, section heads

or part titles? Another example: "Even if *they* are not used in the final book, section head numbers are useful in the outline." This can only read well if it is rewritten to read: "Even if section head numbers are not used, *they* are useful in the outline." The rules for pronouns are given in most grammatical guides, but because they are frequently ignored, they are repeated here: *No pronoun without a principal. The principal should not be very far off. There should be no doubt about which of two parties the pronoun refers to. One pronoun, one job. The pronoun should seldom precede its principal.*

- Shorthand usages and abbreviations should be kept to a minimum. Each field has its own accepted abbreviations, well known by specialists, but definitely not by students. IMF will have a different meaning in the field of economics than it will in the field of physics.
- Erudite abbreviations should be avoided. For instance, why use "e.g." for "for example" or "i.e." for "that is"? Many students will have no familiarity with these abbreviations.
- Don't assume that students know how to read. They generally don't read very well.
- Backward references should be used frequently, but forward references sparingly. When backward references are used, the student will be able to relate the current topic to principles already learned (learning is best accomplished by repetition). Forward references should never be to general concepts, but only to specific topics or illustrations.
- Frequent summarization is desirable, briefly at the ends of sections and more extensively at the ends of chapters. Such summaries should refer back to earlier discussions.
- The author should write directly to the reader. Although the first person should be avoided, it is acceptable to use "we", provided it is used consistently and properly.
- The level of the audience should be kept clearly in mind.
- A consistent style should be maintained. Passive voice should not be used in one sentence and active voice in the next.
- The style of the book should be broken up, with frequent use of #1 and #2 heads.
- Paragraphs should be kept short.
- Occasional, but not too frequent, use should be made of numbered lists. However, such lists should not be too long.
- Words that are being defined should be set off typographically, by

use of bold face, italics, or color (the book designer will make the actual decision, but the author should mark such words).
- Headings and subheadings should be kept parallel in construction (this is where the outline provides great help).
- Sentence structure is important. Complete sentences should always be used but sentences should be kept short. If a sentence has more than one dependent clause, it should be rewritten into two sentences. Judicious use of commas greatly improves the readability of most longer sentences.
- Every paragraph should move the topic forward.
- Attention should be taken to ensure that the book isn't dated by references to current events, specific years, or names of current public figures.
- Bias should be scrupulously avoided. Great attention should be taken to ensure that sexes, minority groups, the elderly, and disabled people are given equal treatment. Some guidelines for bias-free publishing are given in an Appendix to this book.
- Quotations from others should be avoided unless they are absolutely essential (a discussion of fair use and permissions will be covered later).
- Important points should be summarized before new points are begun.
- A concise summary should be provided for each chapter.

Illustrations

The modern textbook is heavily illustrated and is designed to have strong visual appeal to the student. Illustrations are important because they break up the monotony of the text, provide alternate ways of presenting material ("one picture is worth 1000 words"), and help to make the material exciting for the student. Certainly, illustrations are a major factor in the adoption of textbooks. However, illustrations can also markedly increase the cost of producing the book, and can cause antagonism or confusion if they are not done properly.

An illustration program can be handled either by the author, or by the publisher. On many large textbook projects, most or all of the illustration program is handled by the publisher. This greatly simplifies the author's work, but at the same time the author loses control. Photo researchers, artists, and designers hired by the publisher may do an excellent job of illustrating the book, but the potential for professional embarassment exists. I have heard a number of horror stories about books that have really been ruined by less than professional art programs. However, it is also true that most authors have very little idea about what illustrative material will really work in a textbook. The ideal arrangement would be a close collaboration between author and art director, with the author calling all the shots. This arrangement can rarely develop. There are a few highly qualified free-lance art directors who will work closely with an author and publisher on a project, but they are expensive and their cost can only be justified on a book with a very large anticipated sales. If the publisher is assuming responsibility for the art, it should be so specified in the contract, and an additional clause should be added which provides for final approval of all art by the author. I recommend that the author assume responsibility for the illustration program, with an agreed-upon budget provided by the publisher.

There are two kinds of illustrations, line and photographic. These must be handled differently by the printer, and present different problems to the editor and designer. They also have different purposes within the textbook. Line drawings are done by an artist, usually from the author's rough sketches. Photographs are obtained by the author or photo researcher; they may be enlarged or cropped, and labels and overlays may be added.

Line drawings

Although all line drawings are handled similarly by the compositor and printer, from the viewpoint of the textbook author they are not all the same. I have divided line illustrations into several different categories:

- Charts
- Graphs
- Static diagrams
- Flow diagrams
- Maps
- Line art containing chemistry
- Line art containing mathematics
- Artistic renderings (line art containing people, animals, and other familiar objects

Artistic renderings must be done in a much more professional manner than art dealing with abstractions that are unfamiliar to the student.

Graphs and charts are often considered together, because frequently one can illustrate a particular point either way. Sometimes the word "chart" is used in a generic sense to refer to both graphs and charts. To most people, a graph is a line illustration employing X/Y coordinates (Cartesian coordinates). It is generally used to plot the change in a dependent variable (on the Y axis) as a result of a change in the independent variable (on the X axis). Graphs are very frequently used in science and economics, less so in the social sciences, and rarely in the humanities. A graph is often a rather formal looking structure that is unfamiliar to the student. The author's job is to convert the graph into something intelligible for the student unfamiliar with the field.

A chart is often used to illustrate material similar to that in a graph, but the manner of presentation is different. The two common types of charts are the "bar chart" and the "pie chart". The bar chart is generally used to show how several elements differ, without actually showing them in Cartesian coordinates. The data in both charts and graphs can also be presented in tabular form, and it requires a specific decision in each case as to which is the appropriate way of illustrating the material.

Diagrams are line drawings intended to illustrate a process or procedure. They can either be static, illustrating an item or procedure in a single panel, or they can be dynamic, indicating the flow of a process, or a series of steps in a procedure. In general, diagrams are preferable to graphs and charts in a textbook, because they can be made less formal and rigid. A good artist, given the right instructions, can turn out attractive and very useful diagrams that will make any textbook look and sell better.

It is very important to examine the labels on the axes of graphs and bar charts to be certain that they are intelligible to uninitiated students. It is almost never possible to take graphs and bar charts directly from the literature without any relabeling.

In some fields, diagrams and charts are little used, but artistic renderings are common. A book on zoology illustrating the various classes of animals is a good example. Here real things that the student is unfamiliar with are being illustrated, and the only way the information can be transmitted properly is with professionally-rendered artistic drawings. Textbooks in the humanities may also find frequent use for artistic drawings.

Sources and preparation of line drawings

In a textbook with a reasonable budget, all line drawings should be specially drawn, using copy provided by the author. It is a big mistake to copy pre-existing drawings, even if copyright permission can be obtained. There are several good reasons to avoid "lifting" material from other sources:

1 The book should look fresh and new.
2 The material in the other book will likely not be exactly what is wanted. To use it unchanged is simply laziness.
3 The act of creating a drawing for the artist to duplicate will force the author to consider what is really desired in the illustration, and why. Nothing is so bad as a drawing taken from another source that has elements that are foreign to the textbook.
4 The quality of reproduction will be higher for newly created drawings than from those lifted photographically from another source.

Even though new drawings will be created, this does not mean that the author should not refer to existing material. Certainly, the monographs, reviews, and research papers in the field must be consulted for ideas. In many cases, a particular experiment or process is unique and can only be illustrated by using another source. However, the author should examine the drawing carefully and redo it in a form that will be distinctive for the textbook.

The success of an illustration will depend greatly upon how well the author has provided instructions for the artist. Generally, the author will never meet the artist face-to-face and must make pencil sketches that have all of the final elements of the drawing, and these must be done in such a way that the artist can transform them. Many artists will copy virtually directly anything supplied, not changing size, relative proportions, or outline. Of course, too much "artistic license" in a textbook is bad, but the author may well want the artist to use some ingenuity to develop a particular illustration. If so, instructions to this effect should be written on the copy. The standard way of writing instructions to artists and typesetters is to circle anything on the copy that is not to be illustrated or printed.

To prepare copy for the artist properly, the author should have a general idea from the editor of the design of the book. Will the book be printed in two colors (or even more gloriously, four colors)? If so, then drawings can be made to do a lot of things more easily than if only one color is used. But, if so, the author will have to mark on the copy for the artist where the second color is to be used. If this is not done, the art editor or the artist may do it, with possibly disastrous results. Is the book going to have a large, two-column format, or a less formidable one column format, perhaps with wide margins? Where will the illustrations likely be placed, and how large are they likely to be? These matters will influence the size and orientation of the drawings. For the first edition of a major textbook, it is very desirable to have a meeting with the editorial and art director before the illustration program is carried too far, to avoid some of the subsequent problems.

Photographs and other continuous-tone material

The editorial and production staff will refer to photographs as "halftones". They are handled by the printer differently from line drawings. Photographs are illustrations containing continuous tones, with many shades of gray as well as bone white and solid black. The printing process is unable to render the gray tones unless the photograph is converted into a series of tiny dots. Wash drawings and some pencil drawings can also only be reproduced by the halftone process.

A halftone is made by photographing the original through a screen that breaks up the continuous tone into a series of dots. Screens vary in size from 65 to as many as 400 lines to the inch. The coarser screens are used for low-quality printing such as newspapers and the finer screens for high-quality printing. If any photograph in a printed source is examined with a hand-lens or magnifying glass, the halftone dot pattern will be seen. The quality of reproduction of a photograph in the printed material will depend on several factors: 1) The fineness of the screen used to make the halftone; 2) The quality of the paper; 3) How well the platemaker made the halftone; 4) The quality of the original photograph supplied. Little author control can be exercised over the first three items in this list, but the fourth is the author's responsibility.

To have a high quality halftone, a high quality photograph must be used. First, it should be emphasized that a copy cannot be made of a photograph directly from another printed source. This would result in copying an already existing halftone, and when a new halftone was made from it, the two patterns of dots would almost certainly be out of register and the quality would be seriously degraded. Under special conditions or in times of dire necessity, it may be possible to use a pre-existing halftone, but the quality is almost always low, and the final result may be very poor.

Thus, because of the halftone problem, any photograph to be reproduced must be available as an original positive print or negative. For most purposes, a positive print is preferred, but it must be of high quality, without marks, smudges, or writing. If the photograph was seen in a printed source, then an original print must be obtained from the author. Generally, authors are quite cooperative about providing prints, especially if they are told how the photograph will be used and why it is wanted. It should be noted that in most cases, the publisher of the book or journal in which the photograph appeared will not have the original. (The publisher may likely own the copyright, however, so that permission to use it will have to be obtained from the publisher, even if the author provides the print.) When writing the author, the option of providing either a copy of the already published photograph, or a similar unpublished one might be raised. Many authors will have a selection of photographs, only a few of which were used in their article. An unpublished photograph has the advantage that it is not subject to any copyright restrictions of the publisher and will be fresh.

In some fields, photographs must be obtained not from scholarly publications but from conventional news media. There are many photo research companies that maintain files of photographs and can provide a suitable picture, generally for a substantial fee. The author should be certain that fees for the use of such photos are paid by the publisher rather than the author, and that this is specified in the contract.

Photographs may be used unchanged but often they are cropped so that only part of the print is used. The textbook author should make all cropping marks, since otherwise the art editor will probably use the whole photograph just as it stands, reducing it as necessary to fit the space. If special requirements for size are desired, these should be marked as well. The best way of marking a photograph for cropping is on a transparent overlay. The photograph is mounted on stiff cardboard and covered with a sheet of transparent paper. The crop lines and written instructions are marked lightly on the overlay.

If the photographs depict abstract or unfamiliar material, they should be well labeled. Printers have several procedures they can use to add labels to photographs. The labels are generally positioned outside of the main part of the photograph, and lines called "leaders" then point to the area being labeled. If labels must be placed directly within the photograph, these should be marked clearly on the overlay, care being taken that each leader points precisely to the thing being labeled. (These leaders should be carefully checked on the page proofs to be certain that they still point to the proper things.) When labels and leaders are used, it is essential that the overlay used remain in registration with the photograph, and it is necessary to place small registration marks on on both the photograph and overlay.

The approximate position of each figure should be marked on the manuscript, so that both the designer and the editor will know where it is wanted. Figures should appear next to or after the first reference to the figure in the text, never before. However, the precise placement of a figure cannot always be where it would be most effective, since the actual makeup of the book is a complicated process (see Chapter 6). If a piece of text and figure *must* be exactly together, a conspicuous note to this effect should be included on both the manuscript and the figure.

How large an illustration program?

The illustrations may be the deciding factor in the success of a textbook. How extensive should the illustration program be? This should have been discussed with the acquisitions editor, and should preferably be spelled out in the contract. It is important that a clear understanding of the extent of the illustration program be agreed on early in the project. Most publishers will be willing to specify in writing the approximate number of illustrations for the book, since this is an item that must be included in the preliminary book budget. If the book is to be 800 pages, then at least 400–500 illustrations would be reasonable. A few of these illustrations may be full page in length, but most would occupy only a part of a page. An examination of any modern textbook will show that virtually every page has some sort of visual material. Thus, the author should not feel limited in the extent of the illustration program. However, illustrations should *never* be used just to fill up space. Each illustration should have an instructional purpose and should relate closely to accompanying text.

In the original contract, the length of the book may have been specified in terms of number of words or their equivalent. An illustration that occupies a whole page may be considered equivalent to the number of words that would be occupied by text on that page, but the author should be clear as to what the publisher intends.

Tables

Frequently, the same material can be presented as either a figure or a table. All things being equal, the figure is preferable. Tables often have a formidable look, and students find them difficult to read and assimilate. Tables are most successfully used as a way of summarizing material, or to present in compact form a lot of necessary data. However, the main thought of the student when seeing a table often is: "Do I have to memorize this?"

Tables are expensive to set in type and are often difficult to arrange properly. If a table is being taken from another source, it should be examined to determine if the arrangement is the best for the present book. Frequently, tables in scholarly sources show little regard for the reader, and can be greatly improved by rearranging. Tables are usually clearest if they are arranged vertically rather than horizontally. The left-hand column, called the stub, contains the descriptive information, and the data are placed in columns to the right. With special material, a horizontal arrangement may be used, but only if a satisfactory vertical arrangement cannot be done.

If a table is taken directly from another source without alteration (inadvisable), then permission from the copyright owner must be obtained for its use. If just the data from a table are used, and the material is reoriented or recast in another layout, then permission is probably not needed. This is because copyright deals not with facts but with the way in which facts are expressed. If there is some doubt about whether permission is needed, then it is best to obtain it.

There are two kinds of tables, data tables and text tables. A data table is one in which the entries are numbers, or short words that occupy only a line or two. Text tables have entries which are words, phrases, or even short sentences. Typesetting text tables is much different than typesetting data tables. The problem of copyfitting is often difficult, and the arrangement may not be satisfactory. Items that can be squeezed into a column by handwriting may not fit when set in type. Once the table is roughed out, it should be typed to see how the copy fits. If copy fitting is difficult with the typewriter, the typesetter will probably also have trouble.

Every table should have a title which clearly states the purpose. Explanatory material and source (if any) should be placed in a footnote. The units for each column in the table should be clearly stated. If all of the units for the whole table are the same, then it is preferable to indicate the units in the title itself, or in a general footnote to the table, whereas if each column has different units, the units should be stated at the top of each column.

Each table should be numbered, and the numbering system should be consistent with that used for figures. The placement of tables should be clearly marked on the manuscript. As with figures, each table will be placed near or after the place it is first cited, never before.

Readings

A list of selected readings should be included with each chapter. The purpose of the reading list is to help the interested student find out more about a particular subject, or to help the instructor obtain background material for lectures. The reading list should be short and carefully selected. Articles and books cited should be general in nature, rather than related to very specific points. Advanced textbooks in the field should be cited, since they are usually available and are fairly easy for students to read. An article in a scholarly journal is generally not the best reading for the student. Readings should be chosen that provide reviews of a particular subject and which themselves provide good lists of references. Also, readings should be chosen from sources which are widely available. Most institutions where the book is used will not have libraries with extensive research collections, and may have only a few key journals in a given field.

To be most useful, each reading should be briefly annotated. The purpose of the annotation is to indicate to the student and instructor the value of a particular reading. One sentence is all that is needed to set the scope of each entry. Unannotated readings are generally used only if space problems in the book are severe.

To avoid dating the book, the readings should be taken from up-to-date sources. When the book is revised, care should be taken to modernize the reading list. Of course, a few classic references of historical interest may be cited even in a text which deals with very current material.

It is the experience of most instructors that students rarely voluntarily seek out items on the reading list. Even if an item is placed on library reserve, it is rarely used unless a specific assignment is made in it. Thus, the reading list is there primarily for the rare very interested student or for the instructor who needs help preparing lectures. The citations in the reading list should be complete, and abbreviations of journals should be avoided.

Study questions

In some disciplines, study questions at the end of the chapter are commonly provided. Such questions are a chore to prepare, and poor questions may be less value than none at all, but answering questions is one of the best ways of learning. Selected questions may also be presented as part of the text itself, with the answer completely worked out. Such questions within the text are generally called "Examples", and may serve to clarify an abstract concept just discussed. Questions work best in quantitative fields, where computations are a major part of the activity.

Both examples and lists of questions serve an instructional function. Instructors find question lists useful for making up exams, and for ascertaining that they themselves understand the material. Students use the questions to determine how well they have learned the material. In some courses, the students may be assigned specific questions from the book to answer and hand in for grading.

If the questions involve computation and the answer is a simple number, then the answers may well be provided at the back of the book, or even with the questions themselves. One common format is to provide the answers to every other question. For extensive question lists, a separately published answer book is usually prepared, to be supplied gratis to instructors teaching the course. However, more extensive question lists for use by the instructor are commonly provided in a separately published Teacher's Manual.

The question list at the end of the chapter should be viewed not as a testing tool but as a learning aid. The questions should progress from simple to complicated, and each question should be chosen so that it illustrates an important point. If examples are not used in the text itself, it is very useful to provide in the question list worked-out answers for some of the questions, so that the student is able to see how to proceed. Nothing is more frustrating for a student than a question which is completely baffling. It is also useful in a lengthy question list to indicate the section of the chapter from which a particular group of questions was derived. This makes it easier for the student to get started answering the questions.

The primary guide to the construction of question lists will be the author's own experience teaching the course. Other textbooks in the field may provide some help in determining the approach and level, but questions should never be taken from another book, even if they are changed enough so that the copyright law is not violated. Experienced instructors know that writing a good question is a difficult task, and requires as much creativity as writing a good paragraph of text. Each question should focus on a particular point. Frivolous examples and cute names should be avoided, and questions should be direct and to the point. Which of the following is preferable?

> A gardener feels it is taking too long to water a garden with a 1-cm-diameter hose. By what factor will the time be cut if a 1.5-cm-diameter hose is used? Assume nothing else is changed.

> On a hot sunny day, Mary Smith is watering her garden but the water pressure in her city is low and the watering is taking too long. The hose she is using is 1 cm in diameter. By what factor would her time be cut if she changes to a 1.5-cm-diameter hose? Assume nothing else is changed.

The second example might work well in a secondary school text, but would be inappropriate in a college text. It adds some extraneous material that has little to do with the question, and lengthens the question by a whole line.

Appendices

It is frequently useful to provide one or more appendices which contain material relevant to the text but which cannot easily be placed within the text itself. Sometimes an appendix will be used to provide background material that most students should know before beginning the course, but which some students may lack. For instance, a short mathematics appendix for a physics textbook, or a chemistry appendix for a biology textbook. Publishers frequently insist on appendix material of this type if it is the sort of thing found in competing textbooks.

Another use for an appendix is to provide technical details of material discussed in the text which would confuse most students or impede the flow if incorporated into the text. Appendices of this type are often of primary use to the instructor rather than the student.

A third use of appendix material is the provision of detailed tables of data that are needed occasionally for calculations. For instance, statistics textbooks need tables of functions, and business texts may find it useful to have tables of such things as the present value of money. A calculus book will have a table of integrals as an appendix. A history textbook might find it useful to have a complete printing of the Declaration of Independence, or an outline of key dates in the history of the world.

The following are some specific examples of the kinds of material frequently used as appendices to textbooks:

- Tables for converting from one system of units (for example, the metric system) to another

- The International System (SI) of measurement
- Mathematical signs and symbols
- The fundamental constants in physics
- Abbreviations (see below for a discussion of abbreviations)
- Symbols commonly used throughout the text
- Electronic symbols and abbreviations
- Elementary particles in physics
- International graphic symbols
- Outlines of the classification of major groups of living organisms
- Symbols and atomic numbers for the chemical elements
- Tables of properties of elements or compounds in chemistry
- Tables of values for various mathematical functions, such as logarithms
- Tables of statistical functions

It is very important not to overburden the book with appendix material that will not be used. Appendix tables are often expensive to typeset and are difficult to proofread. Most students will have a handbook available that will give detailed tabular material, and most of the general mathematical and statistical functions can be obtained on pocket calculators. An appendix should be included only if the material will be referred to frequently, or if the book is at a level where it may be retained by the students as a reference.

Endpapers

Although in most books endpapers serve no purpose or are merely decorative, in textbooks they often are used for instruction. Perhaps material that is frequently needed can be neatly arranged so that it can be printed within the confines of the two pages of the endpapers. An example would be the Periodic Table in a chemistry textbook. Endpapers should contain only material that is frequently needed by the student for reference, and should consist of material presentable in outline form. Frequently, material in the endpapers could be placed in an appendix, and whether or not the endpapers should be used will depend upon the material.

Glossary

One of the most useful instructional tools in a textbook is the glossary. The glossary, which is usually three or four pages in length, provides short definitions of the whole range of technical terms that are used in the textbook. Students will frequently use the glossary, as will the instructor.

One common reason for having a glossary is that the chapters are not always used in the order in which they were written. Thus, terms that have been defined in earlier chapters may be unfamiliar to students who have not read these chapters. Although the glossary definition may be somewhat abbreviated, it helps orient the student. It is useful to include at the end of each glossary definition the page number on which the term was first defined. Arranged in this way, the glossary acts somewhat as a special index.

Another approach might be to include the definitions of terms as separate items within the index itself. The advantage of this is that one need not duplicate material in both the glossary and index. The disadvantage is that this sort of glossary lacks visibility and may be less frequently used. Also, the definitions will perforce be shorter than they would be in a separate glossary.

The preparation of the index itself is discussed in Chapter 7.

Special manuscript problems with technical material

Many disciplines make extensive use of technical terms or abstract constructions that must be specially handled by the typesetter. Examples would be the extensive use of Greek letters in mathematics, the presentation of formulas and reactions in chemistry, and the various alphabet letters used in many foreign languages. A number of special symbols are used in the business world, such as the "trade mark" and "copyright" symbols. When preparing a manuscript which contains such material, the author must ensure that these terms have been clearly marked so that the proper typesetting is done.

Generally, if a textbook has a large amount of technical material of this kind, the publisher will select a typesetter that is skilled in handling this sort of material. Although this will make the author's job easier, it is still necessary to prepare the manuscript in such a way that it is clear to the typesetter what is required. The standard way of doing this is to insert the unusual material by hand in the proper place in the typed manuscript. However, it is not always clear from handwriting what is actually intended. Greek letters are the most common problems. For any Greek letters that are written into the manuscript, it is desirable to underline them and mark in the margin exactly what is intended.

Thus, one could write "Greek capital Beta" or "Greek lower case lambda". When doing this in the margin, the notation should be circled so that it is clear to the typesetter that it is instruction and not to be set in type.

An amazing variety of special characters are used with the various languages that employ the Latin alphabet. In all, over 80 special Latin characters are used in such languages as French, German, Hungarian, Finnish, Norwegian, Portuguese, Spanish, and Swedish. If one is writing a textbook for one of these languages, one knows how to proceed. The real problem comes with an author writing a textbook in some other field, where one or another of these special characters must be occasionally used. Generally, this will arise in the bibliography, where the names of foreign authors are cited. One approach is to ignore the special characters in the author's name and treat them like the closest English language character. This eliminates the problem of getting the character wrong, and will not confuse students. Since the textbook is not a scholarly monograph, it is not vital that special characters be employed in the bibliography. If the author does desire to use any of these special characters, then it is essential that they be used properly. The Chicago Manual of Style (see bibliography) has an excellent section on setting foreign languages in type and should be consulted by any author who has much text to write in one of these areas.

The typesetter calls these special symbols and foreign language alphabets "special sorts" or "pi characters". They are relatively easy to typeset with modern computerized typesetters, but still present the problem that they cannot be generated by a normal typewriter. This problem can be handled by special coding with microcomputers and some of this coding is discussed in Chapter 8.

Mathematics

Probably the most difficult typesetting is found in the field of mathematics. Mathematicians, of course, are familiar with how mathematics equations and related material are presented in manuscript form, but mathematics is also used by many other disciplines. The following quote from the Chicago Manual of Style is relevant:

> The uninformed author, by exercising poor judgment in selecting notation and by ignoring precepts of good manuscript preparation, can add enormously to the cost of setting mathematical material in type. Book and journal editors are continually faced with this problem and may even in a borderline case have to reject a manuscript for such nonmathematical considerations.

Modern phototypesetters can generate virtually any character needed, and even entirely new characters can be invented. The Chicago Manual of Style has a whole chapter devoted to typesetting mathematics, which the interested author should consult. It is possible to obtain mathematics fonts for some typewriters, and if there is a lot of mathematics, then one of these might be used. As much of the mathematics as possible should be typed. The most convenient way for the author is to leave space in the manuscript for the mathematics display, and then paste it into its proper place after the final manuscript has been typed. If a mathematics typewriter is not used, then the display material must be written in clearly by hand. It is important to allow plenty of space for inserting the display material, and also to leave wider than normal margins because the copy editor will undoubtedly have to add marginal notes to the typesetter.

It is very useful to number displayed equations, since this permits convenient cross-referencing. In textbooks, the preferred numbering system is similar to that used for tables and figures, the chapter number being followed by the equation number. Thus, Equation 12.34 would be the 34th equation numbered in Chapter 12. However, if a large number of equations are used, it may not be desirable to number all of them. Only those equations should be numbered which must be referred to elsewhere, otherwise the extensive list of unused numbers will confuse students and clutter the book.

Abbreviations

Each field has its own range of abbreviations. It is important to avoid any confusion for students, all of whom are entering an unfamiliar field. My preference is to keep abbreviations to a bare minimum, using only those that are so common that no confusion can develop. Style manuals usually provide extensive lists of accepted abbreviations (the Chicago Manual of Style is especially good on scholarly and technical abbreviations).

If an abbreviation is to be used frequently, the standard practice is to spell the word or term out completely the first time it is used in the book, and place the intended abbreviation in parentheses. If an abbreviation is only used once or twice in the book, then it should not be used at all, but spelled out each time. For a complex field where many abbreviations must be used (e.g. chemistry, physics, astronomy), an abbreviation list, either in an appendix or at the front of the book, should be provided. Many professionals have been using abbreviations for so long that they are not even conscious of them. The student, being a neophyte in the field, will almost always need some guidance. Even such commonly used abbreviations as C for degrees Celsius and F for degrees Fahrenheit should probably be explained when used in very elementary textbooks.

Some special abbreviation problems arise and should be considered: personal names and initials (FDR, JFK), titles before names (Dr., Mr., Ms., Lt.), Company names (Inc., Ltd., Corp.), Agencies and organizations (TVA, UNESCO, YMCA), Geographical names (States, U.S., U.K.), directions (NW, ENE), time (A.D., B.P.), months and days of the week (Jan., Tues.), scholarly terms (e.g., i.e., etc.), measures (in., sq.mi., mm), prefixes of metric units (k, c, m), scientific and technical terms (atm, Btu, Cal, cc, HP, ml, pH, rms, RPM), chemical elements (Hg, Au, Ag), biological molecules (DNA, RNA, ATP), commercial terms (#, %, @, ™). Although abbreviations may be listed in a standard dictionary, and hence can be deciphered by the student, it is preferable to avoid abbreviations that are not standard in the field. Some judgement may be needed here. For instance, if chemistry is a likely prerequisite for a biology course, then a biology text may be able to use a number of chemical abbreviations without special definition.

Many publishers have their own house style manuals which prescribe accepted abbreviations, or one of the standard manuals such as the Chicago Manual of Style may be used. The production editor should also prepare a specific style sheet for the book, based on an analysis of the manuscript. The main purpose of this style sheet is to provide guidance to the copy editor concerning accepted practices. The copy editor (see Chapter 6) will then go through the manuscript and make certain that all uses of a particular abbreviation are the same. The author should be provided with a copy of this style sheet for approval before the copy-editing process begins.

Permissions

Permission must be obtained for the use of any material taken directly from a copyrighted source. The copyright law is fairly complex, and it is not always easy to be certain whether permission is necessary. The whole question of copyright is discussed in Chapter 10, and a detailed consideration of the copyright law is given there. One part of the copyright law permits duplication of copyrighted material under the so-called "fair use" provision, which is the general provision under which copyrighted material is duplicated in educational institutions for handing out to students. However, fair use does not apply to use in a textbook, since publishing a textbook is a commercial activity and does not fall under the fair use clause. Thus, permission to use copyrighted material is almost always needed for a textbook, and should be obtained routinely.

When is permission needed? In general, permission is needed if the material from a copyrighted source is being used exactly as it was in that source, or if it is copied without any significant change of position or wording. If someone else's data are taken and made into a new graph or table, permission is probably not needed. If a figure is redrawn in a new style or arrangement, permission is probably not needed. If a copy of a previously published photograph is used (even part of it), permission is needed. If there is some doubt, permission should be obtained. It is always safer to err on the side of unnecessary permissions.

When the contract was signed with the publisher (see Chapter 9), the author probably agreed to obtain all necessary permissions. To request permission, a letter can be written, or a preprinted permissions form used. An example of a preprinted permission form is given here.

The publisher can generally provide a standard permission form, complete with carbon copies, that will have the standard language preferred.

Requesting permission from professional sources is generally straightforward. Rarely do professional societies or even commercial publishers of monographs present any problems. Under rare conditions, a publisher may request payment, usually a nominal fee to cover costs of handling the permissions forms, but it has been my experience that most professional sources provide permission gratis. Trade book publishers, and general trade media may be more difficult to work with. Their material is used for high-profit activities such as television or other mass media, and they may request fairly steep payment. I have paid as much as $100 for the use of a single cartoon from a national magazine, and some fees may be even higher. Prior arrangement should be made with the publisher (written in the contract, see Chapter 9) specifying how the costs of permissions will be handled. Unprompted, most publishers will be happy to let the author pay all of the costs, but may be willing to cover the costs if requested. Some

Sample permission form

Date:____________

____

____

Dear ______________________________

I am preparing a textbook on ____

to be published by Textbook Publishers Inc., on or about ______________________________

My book will be approximately ______ pages and will be case bound. The estimated price is ________. There will be no trade sale.

I request your permission to include the following material in this book:

and in future revisions and editions thereof, including nonexclusive world rights in all languages.

These rights will in no way restrict republication of your material in any form by you or by others authorized by you. Please indicate agreement by signing and returning the enclosed copy of this letter. Should you not control these rights in their entirety, would you kindly let me know to whom else I must write.

Unless you indicate otherwise, I will use the following credit line:

I would greatly appreciate your consent to this request.

Sincerely,

____

____ (address)

____

Agreed to and accepted:

by ________________ ________________ ________________

Signature Title Date

publishers may be willing to provide a specific budget for permissions, thus indicating an upper limit that they will be willing to pay.

For requesting permission from domestic sources, a stamped, self-addressed return envelope should be provided, and one or two copies of the permission request should be enclosed. If a letter is used, a box at the bottom of the page for signing can be included. The author should be specific about what is being requested. If Figure 12.4a on page 99 of volume 53 of the Journal of Anthropology for 1934 is desired, then all this information should be given. The author's name should also be given, if the request is mailed to the publisher. In the request, the following information should be provided:

- The title of the book being written
- The name of the publisher for the book being written
- Whether there will be any trade sale (such as in a general book store)
- A request for nonexclusive rights
- The extent of rights being requested. Nonexclusive rights may be granted just for one edition of the book, for all editions, for just the English language, or for all languages (world rights). It is best to ask for nonexclusive world rights for all editions. Many publishers will grant this, but a few will restrict permission to just the current edition. This means that if a new edition is published, a new permission request will have to be made.

In the upper right hand corner of the permission request form, the figure or table number in which the material will be used should be typed. It is essential to keep a copy of the permission request, and send at least two copies to the publisher from whom the request is being made.

Once the signed permission form is received, a copy should be made and the signed form sent to the author's publisher. The publisher will file it for future reference or in case any question of infringement arises. It is very important to keep copies of all *signed* permission forms, because the filing systems of publishers frequently break down. I had the horrible experience that one of my publishers lost all of the permission forms that had been obtained for the first and second editions of one of my textbooks, many of which had granted exclusive rights for all editions. When beginning production on the fourth edition, the publisher, not being able to find the required permission forms any more, asked me to obtain *new* permissions. This amounted to several hundred permissions, and for some the copyright holder may have been dead or inaccessible, necessitating a lengthy copyright search. Fortunately, I still had my copies of the signed forms and could avoid having to obtain new permissions.

Legally, permission has to be obtained only from the copyright holder. Although copyright rests with the writer, and exists from the moment of creation, in most cases the copyright is assigned by the author to the publisher (see Chapter 10). In the case of unpublished works, the copyright resides with the author. Even for letters, the copyright resides with the writer, not the recipient. With professional journals and books, the copyright holder is generally the publisher, but most publishers will request that permission also be obtained from the author. If there is any doubt as to the identity of the copyright holder, a request to both the publisher and the author should be made. Copyright is only good for a certain period of years, but whether something is or is not under copyright is often difficult to ascertain. The copyright law changed in 1978 and the current term of copyright is longer than formerly. This matter is discussed in Chapter 10, but in general it can be stated that anything published *before* the 20th century is now completely free of copyright restrictions and can be used without permission (but with citation of the source, of course).

Permissions should not be obtained until the final version of the manuscript has been prepared, since changes may occur. Obtaining permissions is a chore, involving lots of correspondence and possibly some expense. For this reason, as well as to impart a unique character to the book, it is preferable to develop new material rather than to use material from others.

Final manuscript preparation and delivery

Once all of the elements for the book are completed, it is time to prepare the final manuscript for delivery to the publisher. Versions of the manuscript that have gone out to reviewers need not be complete, and may be missing certain items such as some figures, readings, problem sets, tables, appendices, and glossary. But the final version, the one that is going to be set in type, should be complete in all materials. The goal at this stage is to prepare a version that can be turned into a book in the most expeditious and error-free manner. Attention to a number of minor points will make this easier both for the author and for the publisher's staff. An example of a manuscript submission checklist is given here.

Everything should be typed double-spaced on 8 1/2 X 11 inch plain white paper. The margins should be kept wide all around, because there will be a lot of marks before the manuscript goes to the typesetter. Only one side of the paper should be used, and a chapter and page number should be placed on each page. If a page is partly blank, a line should be drawn to the bottom to indicate that there is no missing material. If a page has been added after the page numbers have been inserted, it should be designated with a letter. Thus, page 16-5A is a page in Chapter 16 that comes after page 5. If a page has been removed,

MANUSCRIPT SUBMISSION CHECKLIST

TEXTBOOK PUBLISHERS INC. 999 WEST RIVER DRIVE, NEW YORK

Author: You must complete this form and return it with your manuscript

Title ____

Author ____

If more than one author, indicate the author to receive edited manuscript and proof

____

If extensive travel is planned during the next year, please let us know as soon as possible, so that we can arrange the production schedule to avoid conflicts.

Office address of author

____

Phone ____________

Home address of author

____

Phone ____________

Check the elements that are included in your manuscript

Title page ________

Dedication ________

Table of contents________

Foreword________

Preface________

Acknowledgements________

Quotations________

Footnotes________

Questions________

References________

Appendices________

Tables________

Answers________

Glossary________

Bibliography________

Illustrations________

Figure captions________

End papers________

Note! If any element that you've checked is not with the manuscript, or is not complete, please list and indicate when the missing material will be sent.

Will you compile the index to your book, or should we engage a professional indexer and charge the fee against your royalties? Yes ______ No ______

Illustrations

Number of line drawings ______ Number of photographs ______

Please indicate the position of each illustration in the manuscript

Will any illustrations be directly reproduced from copy you supply? Yes______ No______ If so, provide a reproducible copy and not a duplicate, if possible.

Are you planning to use art from previous editions? ______ from other books you have written? ______ or from books or articles by other authors?______

Permissions

Please indicate if your manuscript is accompanied by the following:
Completed permission checklist ____________
All completed permissions forms ____________

Please indicate if your manuscript contains any of the following elements that have been taken from sources for which permission is required:

Quotations________
Tables/charts/maps________
Illustrations________
Readings________
Photographs________
Poetry________
Music________
Song lyrics________
Letters________
Business and/or court cases________

If a signed permissions form is not available for any element checked above, please list the element below and indicate when a signed permissions form is anticipated.

Do any of your photographs contain recognizable faces?________
If so, have you obtained releases from the subjects?________

What type of rights have been granted for permissions? North American________ World in English________ World in all languages________ Please check with your acquisitions editor if there are any questions about rights.

NOTE: Permissions for material being reused in a second or subsequent edition may have to be renegotiated, depending upon the limitations specified. Please check each such permissions form. If there is any question, check with your acquisitions editor.

Copyright Information

Author: In order to register the copyright for your book, the information requested below must be furnished the Library of Congress. Please complete and sign this form.

Title ____
Author ____
Date of birth ________________________ Citizenship ________________________

Year in which creation of this work was completed:
if original work ____________
if revision ____________

For revisions: Give a brief general statement of the material that has been added and in which copyright is claimed.

____

(Signature) (Date)

a note should be added at the bottom of the previous page to indicate what page number follows (at the bottom of page 12-6 the note "Page 12-8 follows" would indicate that page 12-7 had been deleted). It is always more convenient, when putting together the final manuscript, to start each section of a chapter on a new page.

Once the final manuscript has been typed, it should be read again in detail. An amazing number of errors will still be present and are much cheaper to catch at this stage than later. If the errors are minor, it is acceptable to correct them by hand, by inserting or deleting in the body of the text (this is one reason the manuscript was typed double-spaced). Large errors can be corrected by cutting and pasting. As the manuscript is read, the placement of all figures and tables should be marked, so that the editor will know where they are to be placed. When the final book is made up, figures and tables will be placed close to or after the first place they are cited. Also, cross references to other chapters or sections should be checked very carefully, to make certain that they are still correct. Cross references to specific page numbers cannot be made until the final page makeup of the book is completed. At the manuscript stage, such cross references will be indicated by "See page 000". The "000" serves as a flag to the editorial staff that a page number must ultimately be inserted. Cross references to specific page numbers should be done only when absolutely essential, because of the difficulty of getting them correct in the final version. Since the "page 000" will be set in type on the galleys and must be removed at the page proof stage, there is always a chance that an embarrasing cross reference to this 000 page will end up in the printed book. The use of numbered section heads (and even subsection heads in special cases) makes cross referencing almost as precise and much less subject to error (for example, "See Section 9.7").

All material for each chapter should be provided with the final manuscript. This includes reading lists, problem sets, tables, figures (including sketches for drawings to be completed by an artist as well as original photographs to made into halftones), and figure captions. The figures and the figure captions are two separate elements in book production, even though the author will tend to think of them together. I find it useful to write the figure caption on the sketch for each figure that is to go to the artist, but the editor will also want a separate caption list.

Care should be taken to provide all of the material that was specified in the contract. The publisher can (and often will) arrange to have missing material prepared by someone else but then may charge the cost against the author's royalties.

Front matter

Front matter includes such things as title page, dedication, table of contents, foreword (written by another), preface (written by you), and acknowledgments. At some stage, the publisher will require copy for the front matter. A preface may have been written when the first proposal for the book was prepared, but that preface will probably have to be greatly changed because the book has probably greatly changed. The preface is one of the most widely read parts of a book, and instructors thinking of adopting the book will read the preface to look for clues as to intended level and scope. The editorial staff may have a lot to say about what is included in the preface. Even the marketing department may get involved in reading and refining it.

A dedication is not essential, but is usually included. It is a good place to render immortal the names of a spouse and children, or dog, or favorite mentor. Whether or not the dedication will end up being a separate page in the book will depend upon how the final page makeup develops. If there is not enough space for a separate dedication page, the dedication will likely be included on the copyright page.

Back matter

This includes all appendices, glossary, bibliography (if not included at the end of each chapter), answers to questions, and index. If some of this material is not ready at the time the final manuscript is prepared, it can be added later, provided that some indication is made of its existence. The index, of course, must await final typesetting and layout.

Delivery of the manuscript

The acquisitions editor is hopefully anticipating the final delivery of the manuscript. Since a lot of reviewing and editorial work has already been done, the acquisitions editor is probably ready to get the book into production as soon as possible. Scheduling of production for a textbook is a very complex matter. Generally, it takes about a year from the time the final manuscript reaches the publisher until the bound books are ready. Sometimes it may take longer than a year, if special production problems arise or if there are strikes at a major supplier, but it hardly ever takes less than a year. For a book to be examined by professors for fall adoption, it *must* be in their hands no later than March. An earlier availability date is even better. An ideal time for a book to appear is in late November or early December. When appearing at that time, it can have a copyright date for the following year, and a few spring adoptions may be obtained, permitting the publisher to get a little marketing experience. The author should have had preliminary discussions with the acquisitions editor about scheduling, and should try to deliver the manuscript at the proper time. A month or two slippage at this time may mean a loss of almost a year's sales (the author's royalty as well as the publisher's income), and will also make the material in the book effectively a year older.

In most cases, the publisher will only want one copy of the manuscript, the original, and will duplicate it for the many copies needed. The author should, of course, keep a copy. My practice is to make a photocopy of every single page, even those with very little material on them. It is desirable to deliver the final manuscript and all illustrative material in person. This ensures that the material arrives safely, avoiding potential loss in the mails, and also makes it possible for the author to meet production personnel. Since the photographs being provided are probably one of a kind, it would require a tremendous amount of work to replace them if they were lost.

However, a personal visit is only worthwhile if arrangements can be made to meet some of the production staff who will work on the book. Simply dropping off the manuscript at the publisher's front office is of little value. Thus, when the completion date of the manuscript is fairly certain, the acquisitions editor should be contacted and a personal meeting with key production people arranged. The key people to meet include the person who will be doing the editorial supervision, the art director, the designer (if different than the art director), and the person in charge of manufacturing. It might also be possible to meet with the director of marketing. With the manuscript on the table, a very productive meeting can often be held, and a number of potential problems anticipated. If nothing else, such a personal meeting will make less impersonal any disputes that may arise during the complicated production process.

Once the manuscript has been delivered, the author can sit back and relax for a little while. The ball is in the publisher's court, and hopefully it is fielded expertly. In the next chapter, we discuss what happens to the manuscript once it reaches this final stage.

6

From manuscript to printed book: the publisher's perspective

When the final revised version of the manuscript has been prepared incorporating all changes suggested by reviewers, it is time to begin production of the actual book. This is a complex operation involving many people both inside and outside the publisher's organization. The author still plays an important role, but this role becomes progressively smaller as the production process proceeds. In fact, if the author were to drop dead at this point, the publisher would still be able to bring out a successful textbook (albeit with some difficulty).

The final manuscript, incorporating all comments of the reviewers, is submitted by the author to the acquisitions editor, whose job it is now to ensure that the project gets transferred successfully to the book production department. In a large publishing organization, many books are being produced, and each book will have to take its place in a queue. An experienced acquisitions editor may be able to expedite the process of transfer to production, but an inept acquisitions editor can be the cause of many delays. It is not uncommon that a manuscript at this stage sits idle for many months (in rare cases, years!), waiting for someone to give it a shove ahead. In the worst case, the acquisitions editor who signed the book may move to another company. The book then becomes an orphan, waiting for someone to come along and adopt it. It is the author's job (see Chapter 7) to make certain that production begins quickly, since every month lost is lost royalty income as well as another month toward having data and references become out-of-date.

In the present chapter, we will present the steps in the book production process, from initial typesetting of the manuscript to final binding of the book. The discussion of the physical aspects of book production will be necessarily brief, since much of this is handled outside the publisher's organization and is beyond the control of the author. We will discuss in some detail the people who are involved in the production process, and the manner in which these people are brought together to work efficiently on the project. The book production schedule will be covered in some detail, since this is crucial to the success of the project, and impinges directly on the author's activities.

Overview of the book production process

The publishing house almost never has its own printing plant, and the printing of textbooks is almost always done by specialized book printers. Books are now produced almost exclusively by means of offset printing. In this process, raised type is not used; metal printing plates are prepared photographically and wrapped around a drum. There is a difference in wettability between the photographic image and the matrix of the plate, and the ink adheres only to the image. From the metal plate, the ink is transferred to an intermediate roller and from there the image makes an impression on the paper. Because of the intermediate step, the metal plate itself does not have to be pressed against the paper, thus making for a long plate life. The metal printing plates inserted into the press contain the image in a positive form. A single printing plate may have 16 or 32 pages of text, all of which are printed on one side of the sheet in one single impression.

The most common type of press used for printing textbooks is called a "web press". The paper comes in large rolls and is fed into the press in a continuous stream. Both sides of the paper are printed in a single pass through the press. In line with the press are machines which cut from the roll the large sheets of paper containing the printing images, fold them into 32 or 64 page packages called "signatures", and trim off the excess paper. The separate signatures that make up a complete book are then gathered in order and bound inside the cover (which has been produced separately). If colors other than black are also to be printed, then separate printing plates containing the images for these other colors must be prepared. Each color is printed in turn as the paper passes down the length of the web press.

Anything that can be photographed can be converted into plates suitable for printing. The publisher's job is to produce the copy, called "camera-ready copy", that can be properly photographed. Camera-ready copy is prepared for each page of the book and for a single page consists of the typeset manuscript plus all line drawings, tables, figure captions, running heads, and folios (page numbers), arranged on the page in the desired locations, but not photographs.

Photographs or any other elements containing tones other than solid black are called "halftones" (see also Chapter 5). They can only be printed satisfactorily if their images are converted into an array of tiny dots. To make a halftone, the picture is photographed through a screen containing a grid of very fine lines. The number of grid lines per unit area determines the spacing of dots on the halftone photograph. A common screen ruling for textbooks is 133 lines per inch. Finer screen rulings of 200 to 300 lines per inch are used for very high quality halftone reproduction. The sizes of the dots in an area of the photograph depend upon the grayness or blackness of the image from which the halftone has been made. An area which is light gray will have very small dots surrounded by white, whereas a dark gray area will have larger dots surrounded by very little white. The distance between dots is a function of the ruling density of the halftone screen. Because they must be converted into halftones, photographs are handled separately from other material in preparing camera-ready copy, but proper space must be left for them on the page.

Each page of camera-ready copy is then photographed by a large high-quality process camera to produce a negative film of that page. Depending on the sheet size to be used in the printer, the negative may include 16 or 32 pages. All halftone negatives, prepared as just described, are then inserted in proper place on the film, a process called "stripping". Each page is positioned on a large sheet of specially coated masking paper which contains openings in proper position for each page. Pages are of course not in the same order on the mask as they will be in the final book, since even numbered pages are on the back of odd numbered pages and are printed from separate plates. The process of positioning the pages properly is called "imposition" and the final sheet with negatives in place is called a "flat". Once all of the pages are imposed, the images on the large sheet are transferred photographically to the final metal printing plate.

The stripping process and preparation of printing plates is almost always done in the printer's establishment, but the preparation of film negatives and halftones may be done elsewhere, perhaps by the typesetter. The accuracy of imposition must be very high, and this is most critical when more than one color is being printed on a single sheet, since separate printing plates must be prepared for each color. In high quality systems, accurate alignment of all elements is ensured by the use of pin registration, which makes use of special holes and pins on copy, film, plates, and presses. If the pages of a two-color textbook are examined carefully, the accuracy of the registration can be assessed. Registration frequently varies considerably from book to book, and between pages of a single book.

The text for the camera-ready copy is prepared by a typesetter (sometimes called a compositor). The typesetter output usually consists of pages of photographic paper containing the text in positive form. The machine which produces the typeset material is called a "phototypesetter" and is driven by a computer.

Typeset material differs from conventional typewritten material in a number of significant ways. The sizes and shapes of letters are under much more detailed control in a typesetter than in a typewriter. For instance, the space that each letter occupies on a typewritten page is always the same, whereas with a typesetter, the spacing between letters can be varied. Thus, the letter "i", being narrow, occupies less space than the letter "m", and the computer which drives the typesetter is able to make decisions on such matters. The output text, called "proportional spaced", looks much nicer and reads better than typewriter output. The typesetter is also able to "justify" the text, making the right margin of the text align down the page. This is done by varying the spacing between words on each line, and by hyphenation. Justified text generally is considered to look better and to read better, although books are occasionally printed with unjustified text.

To achieve suitable, proper justification, long words falling at the ends of lines must be hypenated. The computer hyphenates by use of a set of word-break rules and by the use of a large exception dictionary which includes those words which do not fit the rules. Hyphenation and justification are completely automatic in most modern typesetting systems, but the result must be checked by a human to be certain that crazy mistakes have not been introduced. For instance, how should the word *resort* be hyphenated? It depends upon the meaning:

The family went to the res-ort

It is time to re-sort the cards

Another major feature of typeset text is the availability of a large number of special characters that can be printed. Characters such as @, #, °, ©, ®, ™, α, β, Π, that are not part of the conventional typewriter character set can be readily printed by the typesetter. Special characters find great use in many technical fields and are widely used in textbooks. Foreign accents can also be readily set by the typesetter: é, Ö, ñ, Å.

The characters of the text are transferred into the memory of the computerized typesetter by keying at a computer terminal, and the computerized phototypesetter then performs the conversion into typeset form. Either the keyboard operator or the computer itself makes decisions about spacing and justification. If the author has a microcomputer or word processor for writing the text, as discussed in Chapter 8, then most of the keyboarding by the typesetter company can be eliminated, provided the typesetter is able to interpret the format codes used. When characters are typed at a computer terminal, they are captured in "machine-readable" form, and any appropriate computer should be able to interpret and convert them. However, the special characters that cannot be typed on a conventional keyboard must be specially coded, and the typesetter's computer is able to interpret these codes.

The output of the typesetter can either be positive or negative photographic material. Most typesetters generate positive material, with black images on white photographic paper. The output can be either a long stream of continuous text, called "galleys", or sheets already converted into page form. For highly illustrated books, galley output is generally used because it is difficult to allow proper space for the figures, but for straight text, it is quite feasible to produce page output directly from the typesetter.

The typesetting can be done either by the printer who will produce the final books, or by a specialized typesetting company. For most textbooks, typesetting is done by specialized typesetting companies that are not connected with printers. In a few cases, the publisher may have its own typesetting facilities.

The copy that is provided to the typesetter is the final edited manuscript containing detailed instructions for design, format, and manner in which technical material is to be handled. This copy has been read thoroughly and marked for the typesetter by the "copyeditor". The copyeditor is a specialized individual, frequently self-employed (free-lance), who works on hire from the publisher for a single job. In rare cases, the copyeditor may be an employee of the publisher. Once the typeset output has been produced, this will also be read by a copy-editor, who will proofread it and mark all errors. The author will also be asked to proofread the material.

While the text of the book is being handled by copyeditor and typesetter, the illustrative material is being prepared by others. The art department of the publisher will be responsible for overall supervision of the art program, hiring free lance artists and supervising their activities. If the publisher is providing photographs for the book, the art department will also contract with photo researchers, whose job is to find suitable photographs. Ultimately, the illustrative material and the textual material are put together to generate the camera-ready copy suitable for conversion into printing plates.

The arrangement of all of the elements on a single page is called "make-up", and is generally done by a book designer (either a free lance or an employee of the publisher). The material to be affixed to the page is treated with a special adhesive wax and is then carefully positioned in the proper place. Final line drawings reduced photographically to proper size are fixed in their desired positions, and running heads, figure legends, folios (page numbers), and other elements (such as footnotes, marginal notes, etc.) are fixed in place. The final page, called a mechanical, will contain all elements except the photographic halftones, which will be stripped in on the film negative that will be made subsequently. Makeup of a heavily illustrated book is a very complex activity. Each page should have the same number of lines as every other page, the break in the text at the bottom of the page must not come at an inappropriate place, and the illustrations must be placed close to or after the place where they are first mentioned. For a heavily illustrated book, before the final mechanical is made, a preliminary version is worked out which is called a "dummy". Using an exact copy of all of the elements, the person doing the dummy works through the whole book page by page. If an inappropriate page break develops, it may be necessary to go back through a number of previous pages, repositioning illustrations or making other modifications, in order to create a suitable set of pages.

One of the things that the person making the dummy pays special attention to is having the book terminate at the end of a signature. As we have noted, books are printed in signatures, generally 32 or 64 pages in length (shorter signatures are possible, but add to the expense of the book). Any pages at the end of the book that are not used constitute just so much wasted white paper (paper constitutes about half the printing cost of the book).

The production editor

We thus see that the production of the book is a complex process that involves the activities of a large number of people, many of whom operate outside of the publisher's organization. The whole operation of producing the book is supervised by the "production editor", who is the key to a successful book. The production editor is almost always employed directly by the publisher. It is the production editor with whom the author generally works directly. The production editor will be the conduit through whom copy, typesetting, art, and all other elements pass to and from the author. The production editor will co-ordinate the whole job, working to maintain the schedule that has been set, cajoling authors or suppliers to get their work done on time, consulting with the acquisitions editor about major changes in the book, feeding material to marketing and sales, etc. With a good production editor, virtually anything is possible and a high quality book is likely. The most common reason for a poorly produced book is a poor production editor. Unfortunately, the author has very little control over the selection of the production editor. The most important thing is to have a production editor who has considerable experience within the company. A relatively new production editor will be at a disadvantage when it comes to getting work out of others, and may find it difficult to arrange things in the way the author would like. An experienced production editor will "know the ropes", and will be able to cut corners, get around intransigent colleagues, anticipate difficulties and prepare for them, and in general drive the book to a successful completion. For a major textbook, the production editor may work almost full time on this one book. In contrast to many others involved in the production of the book, the production editor is generally a full-time employee of the publisher, although smaller publishers may employee free-lance production editors.

It is the production editor's job to see that the schedule is maintained. Although a major textbook can be published in a few months if a crash program is undertaken, it generally takes about a year from the time the book is launched into production until bound books are available in the warehouse. This schedule can be shortened somewhat for smaller books, and may be extended for a very large or complex book, but the year schedule is something that seems to be fixed in the textbook publishing industry.

One way for the author to ensure that a good production editor is on the job is to have this written into the contract. If the publisher is very interested in the project, and perhaps is competing for it with one or two other major publishers, then the publisher may be willing to write into the contract language specifying a senior production editor by name. Some of the large publishers have several separate production departments, with a special production department that does major textbooks. If an author can arrange to have the textbook produced by a special projects department, the chances of a high quality book resulting are increased (although by no means ensured).

The copyeditor

The copyeditor is the individual who reads the manuscript and edits it for style and consistency. The copyeditor will also mark items of special typography and insert other instructions for the typesetter. In most publishing houses, the copyeditor is a free-lance worker, hired for the specific job. Copyeditors often live in places remote from the publisher's location, and the manuscript is passed back and forth by mail. The general flow of manuscript is from production editor to copyeditor to production editor to author. If the schedule is tight, the copyeditor may mail material directly to the author.

Copyeditors are paid either by the hour or by the job. The amount of stylistic editing that the copyeditor will do may depend upon how much money the publisher wants to spend on the book. If the book is to be a major textbook, the publisher may be willing to spend more money having it copyedited, but a run-of-the-mill textbook may have virtually no stylistic editing done. In this latter case, the only copy editing done will be that needed to ensure proper typesetting and makeup.

Copyeditors frequently specialize in subject matter areas. A copyeditor capable of doing technical manuscripts well is often hard to find, and rarely will the copyeditor be intimately familiar with the field of the textbook. Because the job is so important, it is to the author's advantage to have a good copyeditor. The author should insist on one who has expertise in the field. University communities frequently have a number of free-lance copyeditors, and the author may know someone locally who would be able to do the book. The publisher may be willing to hire someone an author recommends (but don't count on it, as this reduces the publisher's control over the project). There are some major advantages of a local copyeditor when a tight production schedule is involved, since material can pass between the copyeditor and author without the intervention of the mails.

The art director

The illustrative material is handled by the staff of the art director. Although the author rarely deals directly with the art director, this individual is usually of equal importance with the production editor on major book projects. Activities that are coordinated by the art director's office are the design of the book and the cover, the preparation of line drawings, location of half-tones, and the preparation of the dummy. The art director's office will also usually be responsible for preparation of the camera-ready copy that goes to the printer. Some of the people under the art director's control will be free-lance, whereas others will be direct employees of the publisher.

The artist

Generally, the artist(s) will be free-lance workers. On a major project, the selection of the artist is a critical matter. The art director will select the artist based on the type of material to be drawn, as each artist has a distinctive style and range of capabilities. For instance, an artist who is excellent for graphs and charts may do a poor job of drawing people.

From the author's viewpoint, there are major advantages of having a local artist, since it is then easy for the author and artist to discuss the author's intentions in any problem figures. Most art directors of major publishers will prefer to work with artists they are familiar with, however, since they will be more certain that a specified schedule will be maintained. A local artist might do excellent work but may be unable to maintain a schedule. However, the identity of the artist is a bargainable item in the contract, so that if the author has strong reasons for wanting to use a local artist, this might be included as part of the original contract agreement.

It is very important that the same style of art be used throughout the book. Thus, a single artist for the whole job is preferable. For this to be possible, it is essential that the artist have plenty of lead time, in order that the production schedule not be delayed. A common reason for delay in the production schedule of a textbook is a tardy artist. Like copyeditors, artists often live in exotic places, where the pressures of commerce do not weigh heavily and where the mails are slow.

The prudent art director will sign a contract with the artist that will specify in detail the number of pieces of art of each kind, when they shall be delivered, and the price to be paid. In general, artists are paid after they have delivered their work in satisfactory form, and the art director can use payment as a means of keeping the artist on schedule.

The general approach on a new textbook is to have the artist first prepare rough sketches of the art, based on the author's drawings, and send these to the author for approval. Each rough sketch will have the layout of the drawing and the labels written by hand. The arrangement of the elements on the sketch will be determined partly by how the final drawing is to be used in the book. If a drawing is relatively uncomplicated, it may be reproduced in fairly small size, perhaps as a marginal figure or in a single column of a two-column book. A complicated drawing will have to be made larger, perhaps even full page. Before the artist has begun work, someone in the art director's office has gone through the author's sketches and has made size decisions, which are indicated to the artist.

Once the rough sketches have been approved by art director and author, the final drawings are made. These will almost always be made at a larger size than they will finally appear in the book, and will be reduced photographically for preparation of the camera-ready copy. For most textbooks, the labels on the art are prepared by the typesetter and pasted in place by the artist. Labels drawn using lettering sets are a thing of the past. For two-color books, there will be at least two sheets for each drawing, one containing the elements to be printed in black, the other for the elements to be printed in the second color. These two sheets will be in registration. Not only must both sheets be proofed, but the way in which the second color has been used must also be evaluated. Generally, the rough sketches will have indicated which elements in the final drawings will be in second color, so that there should be little problem at the final art stage. In books with simple illustrations, rough sketches may not be needed.

Once the final art is approved by all parties, it can then be photographically reduced to the final size, for preparation of the dummy and (ultimately) the camera-ready copy.

Photographic research

Another major activity centered in the art director's office is the selection of photographs. As we have discussed, photographs are handled differently from line drawings, because they must be converted into halftones. For many textbooks, the photographs are supplied by the author, but on nontechnical books, such as those in the humanities and social sciences, the photographs are often provided by the publisher. For such projects, a photo researcher is employed, whose job is to locate suitable photographs. Key sources of photographs are the photo agencies, who maintain stocks of pictures in many different areas, and provide them to publishers on request. The usual way that photo agencies operate is to provide a selection of photos in a particular area, from which the photo researcher and art director will select ones that are suitable. Because of the vast number of available photographs, the main job of the photo researcher is to communicate to the photo agency in considerable detail exactly what is needed. Since often the photo researcher and art director are unsure of exactly what they want, this communication is difficult. The author should have the final veto power on the use of a particular photo, but may find that in the later stages of production and with a tight schedule, approval is difficult to arrange. The author, of course, can also provide photos from local sources, and for many books this is preferable. For technical books in the areas of science and engineering, where photo researchers and art directors generally lack high levels of competence, the author should insist on providing all photos (paid for by the publisher, of course).

Design

One of the main functions of the art director's office is providing the design for the book. The design of a textbook is a complex matter. The position, arrangement, and appearance of every element on the page has to be precisely specified by the book designer. The same text material can be displayed in a wide variety of formats. The impression that the book will make on the user will depend to a major extent on its design.

In most cases, the book designer will be a free-lance individual or firm that specializes in this activity. Some book designers concentrate on the design of books in certain areas, such as chemistry, mathematics, or biology. The design of college textbooks is a specialized field, distinct from the design of trade books or elementary/high school books. Although the author generally never sees the designer, the author should have final approval of the design, and this should be written into the contract.

The size of the final book, after it is cut for binding, is called the "trim size". The trim size is a major item affecting the design of the book, since it determines the width of a text line and how large the illustrations, tables, and other nontextual elements can be displayed. The trim size is influenced by the way in which paper manufacturers and printers function. An odd trim size may lead to a considerable waste in paper and hence an increase in the cost of the book, so that certain trim sizes are more economical than others.

In North America, certain standard paper sizes are available for sheet-fed presses. For books, a common paper size is 25 × 38 inches, which neatly prints and folds into one 32 page signature if the trim size is 6 × 9 inches. Because of this, the 6 × 9 trim size is common in professional books, although it is used only for more advanced level textbooks. The 6 × 9 trim size is difficult to use if large illustrations are needed. Another common paper size is 35 × 45, which is well suited for the production of 32 page signatures in the common trim size of 8.5 × 11 inches. With web-fed presses, other trim sizes are possible.

College textbooks frequently employ trim sizes other than the common 6 × 9 or 8.5 × 11 sizes. Because of the importance of design in the acceptance of a textbook, a considerable amount of experimentation is done. Nonstandard trim sizes can be reasonably economical if the print run is large enough to justify a web-fed press, or, perhaps, a special order from a paper mill of paper in a nonstandard size. Some common trim sizes for college textbooks are 7 × 10, 7.5 × 9, 8 × 9, and 8 × 10.

The decision on trim size is an early decision, because it influences all other aspects of the book design. A second key decision is the width of the text on the page. With an 8-inch book width, two text columns on the page are possible, but with 6- or 7-inch widths, only a single column is convenient to use. The text width has a considerable influence on readability and appearance. A two column format has considerable advantage in terms of flexibility of makeup and location of figures, but has a more formal, "textbooky" look. However, in the larger widths, a single column is not really possible, as the text line will either be so long as to be difficult to read or there will be too much white space on the page. According to many knowledgeable book designers, the most frequent cause of poor readability is excessive length of line. The reasons for this are related to the way the eye moves across the page.

Two factors of textwidth must be considered, the actual length of the line, and the number of characters per line. An ideal line length for a single column book is 4 inches, although many college textbooks use longer lines in order to squeeze more material into a limited number of pages. However, it is not just the length of the line that is important, but also the size of the characters and their spacing, which determine how many characters there are per line. A recommended character count for the 4 inch line is between 55 and 60 characters, although many college textbooks will squeeze more characters into the same space. A single column of 55 to 60 characters per line can be used with a wide trim size, resulting in a wide margin into which many of the figures and small tables can be placed. The flexibility available with modern phototypesetting equipment for setting lines of different character counts is great. A two-column format has the advantage that a shorter line length is possible, and line lengths of 2.75 to 3.5 inches can be used. However, shorter line lengths with fewer characters per line do present problems because frequent hyphenation is necessary to effect justification of the right margin; short lines are thus less preferred by modern designers. The author may have little control over factors related to trim size and text width, except indirectly by how long a book is written. If the book is extremely long and very heavily illustrated, it may be difficult to find a design which does not reduce readability.

The next major factor influencing the appearance and readability of the book is the typeface used. With modern phototypesetters a truly vast array of typefaces are available, and they can be set in many different sizes and spacings. Type specification is generally not something that the average textbook author will want to understand, and is best left to the designer. The author should communicate to the designer (through the production editor) the main concerns about readability, appearance, absence of clutter on the page, and avoidance of trendy or unusual typefaces.

Other aspects of design that influence the appearance and general readability of the book include the way the overall page is laid out, the location, size, and typeface of running heads (or feet) and folios (page numbers), the size, typeface, and arrangement of section heads, paragraph indentations, and footnotes.

Table format is an important design feature in most textbooks. The figure captions and the general way in which figures are laid out on the page also influence the appearance of the book.

The following quotation, taken from a noted book designer, is perhaps relevant to the whole question of book design:

> All innovation and originality implies a departure from a norm . . . .If there is nothing from which to depart, it makes it hard to depart. What is it, then, from which the innovative book designer departs . . . The designer departs, of course, from the book that looks like every other book, the "invisible book," the book that the reading public assumes does not need to be designed, the book that looks just like "a book." It is not a beautiful book, and it is not an ugly book. It does not look cheap or expensive, big or small, plain or fancy. It is completely invisible as a graphic arrangement; nothing comes through but the words and pictures, the author and the images. In fact, when it is described in these terms there will be those who say that this is the ideal book. The most evolved book, like the most evolved Zen master, is perhaps not to be distinguished from the most ordinary. Stanley Rice, Book Design Systematic Aspects, R.R. Bowker, New York.

One problem the author must face is that book designers have egos like everyone else. There are also regional and national competitions for excellence in book design, and the judges at these competitions are not authors or readers but other book designers. A book designer who wins a few of these awards has an enhanced reputation, which makes it easier to obtain contracts for the design of other books. Fads come and go in book design, even at the college textbook level. The designer of the book should be made to understand that the goals of the book are to provide students with readable and interesting information in the most attractive way possible. The author, who should have ultimate approval of the design, must firmly resist any attempt by the designer to construct a book that does violence to these goals.

On any major college textbook, sample pages will be prepared using the designer's specifications. The sample pages will contain representatives of all of the elements in the book, arranged exactly as they would appear in the final printed book. If two- or four-color printing is to be used, the sample pages should be printed in these colors. If a second color is used, several different ones may be tried on different batches of sample pages, in order to see how they look. Approval of the sample pages by the author should be mandatory (and written into the contract). If there are things in the design that are distasteful or unacceptable, the author should make these things known. Frequently, the publisher's staff will resist author's suggestions at this stage, implying that production is too far along to make any major changes.

This is nonsense, as the whole point of sample pages is to provide a preliminary look at the book, from which changes and refinements are to be made. If author approval of design has been specified in the contract, then the author should be required to approve the sample pages in writing before the project proceeds further.

The cover

The cover is an important element in book design that is considered to be of great importance by the publisher. On a major textbook project, even vice presidents may be involved in the final approval of the cover (probably the only part of the book such distinguished managers will ever see). The cover design is frequently carried through into the advertising program, so that marketing and sales departments also get involved. The author should be prepared to give guidance on the cover design, to suggest illustrations that might possibly be used, and to insist on final approval.

The typesetter

Once the design is fixed, typesetting of the book can begin. The manuscript pages have already been copyedited and can now be released to the typesetter for the lengthy and important job of turning the typescript manuscript into characters suitable for printing.

Rarely does the publisher have its own typesetting facility. Although most printers have in-house typesetting capabilities, these are also rarely used, the bulk of typesetting for college textbooks being done by specialized typesetters. There are thousands of commercial typesetters, but college textbook publishers generally employ a relatively few firms who they know from experience can do high quality work and turn it out on schedule.

If the author has prepared the manuscript on a computer or word processor, the bulk of the typesetting has already been done, since the text has already been captured in machine-readable form. We discuss the use of computers and word processors in Chapter 8.

If the typesetter is entering the text at the keyboard, the copyedited manuscript, which has been approved by the author and production editor, is copied character by character into the typesetter's computer. At the same time, the relevant typesetting codes for changes in fonts (bold face, italic, etc.), special characters, section heads, etc., are entered. The output from the typesetter generally will be positive photographic paper (called "repro") or film which contains the typeset characters in the way they will appear in the final book. For most textbooks, the output will be in long sheets called galleys, in which only the main textual material will appear. Page numbers, footnotes, and other elements of the page will not appear in place on the galleys. Major section heads (#1 heads) may or may not appear on the galleys, depending upon the design of the book, but minor section heads (#2 and #3 heads) will likely appear, since they are usually run in with the text. The typesetter's staff will proofread the output and correct any errors found.

Copies of the galleys will be sent to the production editor for checking, and from the production editor or copyeditor to the author. Errors made by the typesetter are called "printer's errors" and are abbreviated "PE"; the cost of their correction will not be charged to the publisher. The galleys that the author sees will likely be marked with any errors caught by the copyeditor. Also marked on these galleys will be the locations for figures and tables, section heads, footnotes, and any other elements that are typeset separately and must be inserted at the final page makeup stage. The author should receive two sets of the galleys, one of which (the master set) has already been marked. It is on the master set of galleys that the author's corrections are to be made. When the author has completed proofreading (see Chapter 7), all of the errors are marked on the master set, which is sent back to the production editor, and on the author's own set, which is kept for reference and perhaps for use in indexing.

Sometimes authors attempt to make major changes in the text at the galley proof stage. This causes the production editor and the typesetter considerable difficulty, because of the expense and complications involved in resetting corrected material. The author will be billed for excess corrections over and above an allowance that has been specified in the contract (see Chapter 9). However, it is understandable that the author may still wish to make a number of such corrections. This is the first time that the material has been seen by the author in type, and typeset material frequently looks and reads differently than manuscript. Misspelled words, improper syntax, and other problems may be seen much more readily in galleys than in manuscript, especially when it is considered that the manuscript has generally been heavily marked up by the copy editor at the time the author last saw it. Another reason for making changes at galley stage is because of changes in knowledge in the discipline. Most college textbooks are written in disciplines that change relatively rapidly, and ideas that were thought to be correct at the time the manuscript was first written may no longer be correct. The author has a perfect right to make changes at the galley stage, and should do so. It is much preferable to make changes at this stage than later, when page proofs are read.

Page makeup

Once all of the corrections have been made, and the galleys have been reset, it is then possible to proceed to page makeup. In parallel with the typesetting activities have been the activities related to the illustration program: line drawings and photos. It is generally possible to keep these two parts of the project in phase, so that they are completed at about the same time.

The next stage, and one of the most critical, is the production of the book dummy. As we have noted, the dummy is a guide to the arrangement of the final page makeup of the book, and all of the required page elements are precisely placed. The dummy is generally done by the book designer, and is a relatively time-consuming operation. It is at this stage that the designer has to cope with all of the peculiarities that the author has inserted into the text. Consider the common problem of a run of pages without any illustrations or tables, and then four or five illustrations in a row, all of which the author wants near the text. Since this is physically impossible, the designer has to move pieces of text and other elements around to find an arrangement that does not do violence to the material yet still keeps the desired elements close together. Since it is poor design to have a figure or table on a page before it is first cited, the designer has to squeeze and fit to make tables and figures fit on pages following their citations. At this stage, it may be necessary to change the sizes of figures, reset tables in smaller form, or completely rewrite tables to make them fit better. The dummy generally results in a long series of compromises, some of which will simply not be desirable, but which cannot be avoided. Some of the major problems that crop up and which end up in print in the first edition may be corrected in subsequent editions, when the author has the advantage of having the first edition available for a guide in rewriting.

The dummy serves as the template which the individual doing the page makeup uses in the preparation of final pages. In contrast to the dummy, the page makeup is mainly a mechanical operation and can be done by relatively unknowledgeable, albeit highly skilled, people.

Page makeup may be done by the typesetter, by staff in the book designer's organization, or by staff in the art director's office. For single-color books, page makeup is generally done with positive typeset material. The pieces of original typesetting, which have not been touched until now except to make copies, are carefully cut apart into panels of the size specified by the dummy, and are arranged on a template, or layout sheet, that has been constructed to the trim size and format of the book. Line drawings have been reduced photographically onto comparable sheets of positive paper, and these are also fixed in place, together with their captions. Other elements, such as tables, footnotes, running heads, and folios, are fixed precisely. Halftones will not be inserted, but the space for each halftone will be carefully marked, so that it can be stripped in by the printer. The final result is a page of camera-ready copy, suitable for conversion into film negative by the printer or typesetter. The halftones are stripped in place on the film negative and the page of film is properly imposed on the large mask used for making the printing plates (as discussed elsewhere).

Two-color books are more complicated, since two separate sheets must be used for each page. With two color books, page makeup is often done on film positive instead of positive paper, because it is essential that the two pages be in exact register. Except for this difference, the page makeup process is the same.

Once the page makeup has been finished, page proofs are prepared which are sent to the production editor and author for checking. This is the most exciting stage of book production, because finally the book is seen almost as it will appear in print. Mistakes show up startingly well at this stage, but are very difficult and costly to correct. We discuss correcting page proofs from the author's viewpoint in Chapter 7, but note here that no changes can be made that will affect the spacing on the pages. If a line must be rewritten, perhaps to correct an error or misspelling, it is essential that the new line occupy exactly as much space as the old line. Production editors and designers are very reluctant to permit authors to try to rewrite the book at page proof stage!

Printing the book

Once the page proofs have been approved, the book can be released for printing. The camera-ready copy is converted into film negatives, one negative corresponding to each page of the book. The production of film negatives may be done by either the typesetter or the printer.

Halftone negatives, which have been made separately to the proper size (as specified by the designer) are stripped (fastened) into place. The film negatives are then imposed in proper order into openings in a large sheet called a "flat" or "goldenrod" (because of the color of the paper used). Final proofs are made at this stage, primarily to ascertain that pages have been imposed in the proper order on the flat. These proofs are made by a blueprinting process and are generally called "blues". The blues are generally not seen by the author, but are carefully checked by the production editor (and perhaps the designer). Once the blues are approved, the printing process can proceed.

Press plates are made from the flats by means of a photographic process. The press plates are generally composed of an ultraviolet-sensitized aluminum alloy. The flat and the metal plate are placed together in a large frame and the images exposed with ultraviolet radiation. The press plates are then assembled and transferred to the press for the printing process. As we have noted, paper for printing is either used in sheets or rolls, depending on the nature of the printing process and the size of the print run. Most college textbooks will have large enough print runs so that web-fed presses using paper rolls will be used. If the book is to be printed in two colors or in full color, then web presses are always used.

The printing of a long run of a two or four color book is an expensive process. In a quality project, the production editor should actually go to the printer's establishment on the day that the book is scheduled to be printed and check the first sheets that come off the press to be certain that the colors are as specified and that other critical features of the book such as halftones are reproduced properly. It is too late to make any changes once the book has been printed.

Once printed, the large sheets are folded into signatures (generally 32 or 64 pages, although smaller signatures may be used to fill out the book to proper length). These signatures are then gathered into sequences and delivered to the bindery. In the standard high-quality college textbook, the signatures for a single book are sewn into a block called a "book block", to which end papers and cover are then fastened by adhesive. Sometimes, the signatures for a hardcover book are not sewn but are glued together by an adhesive, a process called "perfect binding" (which is by no means perfect!). If a paper-covered book is being produced, perfect binding is generally used, although occasionally a paper-covered book is produced with sewn signatures. Other types of binding, more suitable for workbooks and laboratory manuals, include spiral metal, plastic strip, and looseleaf binding. If the author has any control over the design of the book, sewn binding should be insisted on, as this results in a higher quality product. Sewn books open more easily and do not lose individual pages as a result of frequent use.

The covers for the book are produced in a separate operation and are brought together with the book blocks for the final binding process. The cover is fastened to the book block by means of the end sheets and adhesive along the spine. The end sheets may be a decorative item of the book, and can even have material printed on them that might serve as ready reference (see Chapter 5). The actual binding process may be done by the printer, or by a specialized bindery firm, depending upon the nature of the binding and the capabilities of the printer.

Once the books are bound, they are placed in boxes or loaded on skids, wrapped in plastic, and shipped to the publisher's warehouse. The production editor generally obtains a few copies of the finished book immediately from the binder and sends one to the author. The rest of the copies for the author are generally delivered somewhat later, directly from the publisher's warehouse.

The arrival at the author's office of the finished book is hopefully an occasion for celebration. Now comes the lengthy process of marketing the book and obtaining adoptions. Still later will come the book reviews, which hopefully will be favorable.

Budgeting the textbook

A major operation in book production is budgeting, which we have passed over in our discussion of the manufacturing process. The cost of producing a major college textbook is large, and no publisher will enter into a new project without having developed a budget. The budget process occurs in several stages. A rough budget may be developed even before the contract is signed, to be certain that the book proposed by the author will be profitable.

The first detailed budget will be developed once the final manuscript is received. Based on reviewers' comments, the likely market for the book will be determined. Two major factors that must be approximately determined at this stage are the likely price that the book can sell for, and the number of copies that might be sold within a reasonable length of time. Major publishers have extensive marketing experience, and have a good idea of the number of copies that should be sold at different levels and for different college courses. They also know what price books like this are selling for at the bookstore, and what books their competitors have on the market in this area. A final factor in deciding the print run is a consideration of the length of time over which the first printing should be sold out. Although some publishers use periods of two, three, or more years, the standard practice for the major textbook publishers is to print only enough books to sell within the first year. A break-even analysis must determine that sufficient income can be generated from these first-year sales to cover all of the production expenses.

Let us assume that the market for the textbook suggests that about 25,000 copies can be sold in the first year, at a list price of $25.00. The publisher actually only receives part of the price of each book, since the book store must receive a discount, which on textbooks is frequently 25%. Thus, the net proceeds to the publisher from the sale of this first printing are about $450,000, a tidy sum. However, from these proceeds, all of the expenses of producing and selling the book must be paid. This includes all of the editorial, production, and manufacturing expense, the cost of marketing and advertising, the cost of the sales department, and (last but not least) the author's royalty. The only item on this list that is exactly known at the beginning is the author's royalty, which is specified in the contract, and generally will be about 15% of the net proceeds. (The author's royalty on the first printing will thus be $67,500.)

The costs can be placed in several major categories. Initially, costs can be considered as either fixed costs or variable costs. *Fixed costs* are those that are the same whether 1000 or 25,000 books are printed, and include such items as editorial, design, art, typesetting, and all of the other activities that occur before the presses begin to roll. Included in the fixed cost category are acquisitions costs (the acquisitions editor's salary and travel expenses, for instance), and general overhead related to operation of the publisher's editorial and production departments (costs of rent, telephones, postage, and all other items not easily allocatable to a single book project). Also included in the fixed cost category are the marketing and advertising costs, which do not depend directly on how many copies of the book are printed (the actual budget for marketing and advertising will depend importantly on that anticipated first-year sales, however).

Variable costs include all those costs that are a function of how many books are printed. Included in variable costs are the paper, the presswork, cover, and binding. Warehousing costs would also be included in this category. Another set of variable costs are those related not to how many books are printed but to how many books are actually sold. Included in this category are all the costs related to filling orders (invoicing, packaging, shipping). Also included in this latter category is the author's royalty. Finally, there are overhead costs associated with filling orders that are not allocatable to individual books but are included in a general overhead item.

The fixed cost items can generally be estimated in an approximate way from the length and complexity of the manuscript. Depending upon how technical the topic is, the design, art, copyediting, typesetting, and makeup will vary considerably in cost, and an approximate budget will have to be made for these items. Once the book is entered into production, actual costs can be obtained for most of these items. The artist, copyeditor, typesetter, etc., will make exact bids on how much they will charge to do each job, and these bids then become fairly firm and can be used to replace the estimates originally put in the budget.

As the production sequence gets closer to the actual printing process, the cost of production becomes more and more precisely defined. In the early stages of production, only an approximate print run may have been determined, but now that the costs are better known (and perhaps the market has been more precisely defined by the marketing department), a final decision on the size of the print run can be made. The printer doing the job will provide an exact quote for print runs of different sizes, and at this point the final budget decisions can be made.

The final price of the book can now be determined, and this price will be set to provide the maximum profitability on the project, consistent with the prices being charged by competitors for similar books. College textbook pricing is somewhat different than pricing of other books, since the person making the selection decision, the professor, is not the person making the purchase. However, most professors are sensitive to the prices of books, and will certainly hear from students if the books they require are too expensive. Although the author may well be disappointed if the price of the book is high, it should be understood that this price is based on cost and overhead factors that have to be covered. Although college textbook publishing can be profitable, the percentage of the gross sales which ends up as profit is only a small part of the price of a textbook.

According to the Association of American Publishers (AAP), the distribution of the costs for college textbooks can be broken down approximately as follows:

- Editorial, including acquisitions and production; 6%
- Manufacturing, including typesetting, art, printing, paper, and binding; 27%
- Marketing, including advertisements, brochures, catalogs, sales representatives, and complimentary copies; 14%
- Publisher's overhead and order fulfillment; 18%
- Author's royalties; 14%
- Taxes; 10%
- Profit; 11%

Note that all of the percentages listed are averages over the whole college textbook publishing industry. Clearly, individual books will vary from the average, as will individual publishers. Profit, for instance, varies from about 15%, for the largest publishers, through 5% for medium-sized publishers, to virtually zero for some of the smallest publishers. According to the AAP, the heightened profitability of large publishers is a reflection of the economies of scale.

Author's costs and publisher's costs

Although we discuss contracts and royalties elsewhere, it is pertinent at this point in our discussion of budgeting to note that some of the production costs of the book may be borne, sometimes unwittingly, by the author rather than the publisher. Included in production costs that the author must pay are all of the postage, photocopy charges, and telephone bills incurred as a result of producing the book. Frequently, tight production schedules require expensive postage bills for air express or special delivery. Costs for making or copying photographs, as well as secretarial work and rental or purchase of typewriters or word processors are also author's expenses. Although generally paid initially by the author, these are production costs, and should be allocated to the book budget rather than to the author. The publisher would not expect in-house production or acquisitions editors to pay for their own telephone bills or copy charges. Outside suppliers have these charges built into the price they charge the publisher for their services. The author should certainly also be compensated for out-of-pocket expenses. A general principle that should be followed in deciding how author's costs should be allocated is the following: the author's time spent in *writing* the book should be compensated by the royalty, but the author's expenses in *producing* the book should be paid from the production budget for the book. This seems like a reasonable principle, and most publishers will probably agree. Certainly, they should be willing to write language into the contract providing for a certain upper dollar figure for the author's costs for producing the book. Although this dollar figure, like everything else in the contract, is negotiable, $5000 would not be unreasonable for a major textbook project.

Summary

In this chapter, we have presented the procedures involved in the publication of a college textbook from the publisher's point of view. We have indicated the large number of individuals involved, and the many steps at which careful attention is required. Most of the individuals involved in the production of the book are not employees of the publisher, but are outside firms and free-lance individuals who are hired for specific parts of the job. The key individual within the publisher's organization is the production editor, who will be the author's contact inside the company and will be the main channel of communication with the outside suppliers. The success of a book may depend considerably upon the abilities of the production editor.

The costs involved in producing a book are many, and the manner in which they are budgeted and allocated may have considerable influence on the final price of the book. Profit is certainly a small portion of the final price of any book, although how profit is defined may depend upon the accounting procedures of the publisher. The author's royalty is a significant portion of the cost of the book, although if the author's royalty were waived, it is uncertain how much this would reduce the price of the book. The efficiency of the production process may also influence the final cost of the book, although only major failures in book production would likely have any significant impact on the price of the book.

In the present chapter, we have discussed the production process primarily from the viewpoint of the publisher. We have occasionally noted matters that the author should attend to, but we have not gone into author activities in any detail. It is the purpose of the next chapter to discuss the special duties that the author has during the book production process. Once the manuscript is complete, the publisher *could* carry out the whole production process without assistance from the author, but the success of a textbook will depend to a considerable extent upon the author's efforts during the book production process.

7

Keeping the errors out: the author's role in the book production process

In the last chapter we discussed book production from the publisher's viewpoint. We described the various steps in book production and manufacture, and showed how these are handled by the publisher's staff and by outside individuals and organizations hired by the publisher. We have noted the central role of the production editor as a coordinator of the whole process. The production editor is the author's contact during the production process, and is the person through whom all material is passed to and from the author.

In the present chapter, we describe the tasks which the author will perform during the production process, and how these tasks are efficiently accomplished. We will show how the author's activities mesh with the publisher's production activities. The success of the textbook will depend to a considerable extent on how well the author handles the myriad matters that will come up during the production process. By understanding the process and anticipating difficulties, the author's efforts can go a long way toward ensuring a high quality product.

Given ideal conditions, with highly capable production people and the absence of accidents, a high quality textbook may result from a completed manuscript with little effort on the part of the author. However, the author cannot be assured that all of the people involved in the process are highly capable, and accidents do occur. Thus, one of the important goals of this chapter is to show places where things might go wrong, and to indicate how the author can help to prevent disaster.

The work described in this chapter begins after the manuscript has been completely reviewed and all suggested changes have been made. Depending upon how devastating the reviews were, the publisher may have the final revised version reviewed again, to be certain that the problems indicated by the reviewers have been taken care of properly. But ultimately, a final version of the manuscript will be prepared which will be accepted for publication. Acceptance for publication is a major step, because it is at this point that the decision is made (usually by the acquisitions editor and his or her supervisors) to invest a significant amount of money in the book. Since as much as $500,000 may be invested in the production of a major textbook, the decision to publish is not trivial. If a book is to be considered for fall adoption, it should appear sometime between late November and early March, with late May as the absolutely final time. It takes at least a year, frequently 15 months, to turn the manuscript into a book, so that the production process should ideally begin in September. The author can help to ensure that production begins at an optimal time by having the completed manuscript ready at that time. A few months' delay may be critical, and may lead to a year's delay in the production process.

The author should be in frequent communication with the acquisitions editor during the final stages of manuscript preparation, to determine what production schedule is visualized. Although the acquisitions editor may not give a specific date for the start of production, an approximate date should be available. If the acquisitions editor has done a competent job, discussion of a schedule with the production department should occur when the acquisitions editor is fairly certain that the final manuscript is about to be submitted. The author should let the acquisitions editor know that no delays in production will be tolerated and in every way indicate the importance of an early start on the production process.

What should be done after the final manuscript is submitted?

After the final manuscript is submitted, the author should not simply sit back and wait for the publisher to act. Horror stories abound of manuscripts that have been put on the back shelf, or simply forgotten, and have never entered the publication process on an appropriate schedule. Once the final manuscript is completed, it is vital to turn it into a book as soon as possible. Every year lost is so much lost royalty, and since the discipline is moving ahead, the book loses its freshness.

After the manuscript is mailed, the acquisitions editor should be called on the telephone and informed that the work is on the way. The production schedule should be discussed at that time. Once the manuscript is submitted, an acknowledgement of its receipt should be made in writing (inclusion of a self-addressed stamped postcard with the manuscript is advisable). Acknowledgement of manuscript receipt is not only a good general practice, but it may have legal consequences if the publisher has provided an advance or bonus available on completion of the manuscript by a certain date.

After several weeks, some response should have been received from the acquisitions editor about a production schedule. If none has been received, the acquisitions editor should be telephoned and the production schedule discussed. The author should determine whether the acquisitions editor is enthusiastic about the manuscript and is prepared to give it a high priority, or whether it is viewed as one of those run-of-the-mill manuscripts that will be put through a low-budget production process. (Some of these matters may have already been spelled out in the contract.)

The critical point in getting the manuscript into production is a meeting that the acquisitions editor will have with production and marketing departments to discuss the particular project. It is at this meeting that the major decisions are made about the production of the book. A preliminary budget will be made at this time, which will be based on projected sales (see Chapter 6). With a high anticipated sales, a large production budget can be justified. It may not be until this point that it will be decided what kind of production job to give the book: high quality, moderate, low budget. The author should be in frequent contact with the acquisitions editor, keeping the project moving forward.

One of the first things that will be done is a cast-off of the manuscript. Casting-off is the process of estimating the length of the manuscript and from this the length of the book. An accurate cast-off has to take into consideration not only the length of the text, but the number and approximate size of figures and tables, and any other elements that will occupy space.

Once the book has been launched into production, the author should be informed by the acquisitions editor of decisions that have been made. The most important piece of information that the acquisitions editor can provide to the author after the launch is the number of copies anticipated for the first printing. If this number is less than 10,000, gloom prevails. A first printing around 25,000 is excellent, and a first printing of 50,000 or over is equivalent to a jackpot. The author should discuss this first printing in some detail with the acquisitions editor, trying to find out what the basis for the decision was. Almost certainly, the marketing department has had a major input in this decision, but how accurate their forecast is may be questioned. Note that the size of the first printing is usually an estimate of anticipated sales over a one year period.

Once the book is launched, a production editor should be assigned and a production schedule developed. Depending on the budget (influenced again by the size of that first printing), and the publisher, various approaches to production are possible. Small publishers often have all of the production done by free lancers, whereas large publishers will use in-house production editors. Some of the larger publishers have several categories of production, depending on anticipated sales. The author should try to find out if the book is in the top production category. Discussions on some of these production matters were probably held a long time ago, when the contract was signed, and the author should discuss with the acquisitions editor any production items that have been spelled out in the contract (the present acquisitions editor may not have been the one who was involved in the signing of the contract).

The author will know things are on the way when a production editor has been assigned to the book. Once the name of the production editor is known, this person should be contacted on the telephone for a discussion of the production process for the book. If distances involved are not too great, a personal visit is desirable. The author will be working with the production editor a lot, and the establishment of a friendly relationship is valuable. The production editor should provide the author with a tentative production schedule, which should include the dates by which production tasks of the author will have to be completed.

If the author is planning any traveling, this should be discussed in detail with the production editor, as author unavailability is one of the most vexing problems facing the production staff. Hopefully, the author will not be out of the country for any extended trips during the production period. Foreign mails are unpredictable, and rapid air freight service may not be available.

The copyedited manuscript

We have discussed design and copyediting matters in Chapter 6. From the author's viewpoint, the critical thing here is a careful reading of the copyedited manuscript. There are two kinds of editing which the copyeditor will have carried out, mechanical editing and substantive editing. Mechanical editing includes matters such as spelling, capitalization, hyphenation, syntax, quotation and parenthesis marks, figure and table numbers, cross references, and abbreviations.

For each manuscript, a style sheet is prepared which provides specific guidance to the copyeditor on these points. The style sheet shows how various copyediting options should be handled, and tries to anticipate all problems that the copyeditor might have. The style sheet will include special terms and abbreviations in the particular discipline, and how they are to be handled. The style sheet should be shown to the author for approval before the copyediting process begins. Some publishers have a house style and certain standard ways of doing things, but in college textbook publishing, the style to be used will generally also include usages specific to the academic discipline. Most academic disciplines have standard styles that are used by the professional organizations and publications, and unless there are some overriding reasons, the style of the discipline should be used. The author is advised to go over the style sheet carefully, noting each item and relating its style to the preferred style of the discipline.

Substantive editing involves rewriting and reorganizing, in an attempt to make the author's meaning clearer to the reader. Even the most experienced authors can profit from substantive copy editing, but in the college textbook field this is rarely done, as neither time nor money is available. Unless there is some very obvious problem, the copyeditor will probably let the author's words stand. Substantive copy editing in the college textbook field can only be done successfully by someone knowledgeable in the field. Some of this may perhaps have been done by reviewers during the preparation of early drafts. An author sensitive to the nuances of style should arrange for substantive reading by a colleague known to be capable in this task.

Most authors will already be familiar with reading copyedited manuscript as a result of previous publication in professional journals. A number of standard conventions are used in copyediting, and the author should have some familiarity with them. The author should also have been provided by the production editor with an explanation of the typographic conventions to be used for such things as section heads, italics, bold face, etc. It will be the author's job to check the copyeditor's work to be certain that things are properly marked.

An important point: It is easy and inexpensive to make changes at the copyediting stage, but difficult and expensive after the manuscript has been set in type. Any changes that the author makes after having approved the copyedited manuscript may be chargeable to the author's account. The copyediting stage is a good stage at which the manuscript can be read again for sense and nonsense. The author should not hesitate to make any changes deemed necessary at this stage.

Problems or queries to the author will be flagged on the manuscript. These will generally be written on small slips of colored paper that are glued to the manuscript. Each of these queries should be answered by the author, but the slips should never be removed. When a query has been answered or attended to, a line should be drawn through it to indicate that it has been seen. The production editor will go through the author's revisions, will check each slip, and will remove the slip once it has been ascertained that the query has been properly handled.

There may also be colored slips fastened to the manuscript which provide instructions or queries to the production editor or typesetter. The following abbreviations are used: AU, author; ED, editor; COMP, typesetter (compositor).

Any note that the author wishes to write to the editor should either be written on a colored slip of paper (using a color different from that already used) or in the margin of the manuscript. If a note is written on the manuscript itself, circle it to indicate that it contains material that is not to be set into type. All words written on the manuscript which are not circled will be set into type. It is standard practice for each person working on a manuscript to use a different colored pencil. Once the copyedited manuscript has been reviewed by the author, it is returned to the production editor, who goes over the manuscript completely and notes any problems that need settling. Unusual problems may require a letter or telephone call to the author, but in general, from this point the manuscript can go to the typesetter. At this point, the copyedited manuscript represents a major investment of time and money, and there is only one usable copy, the one that contains all of the editorial marks. This copy should be guarded carefully by the author and all others using it, since a lost manuscript will be costly and troublesome to replace.

Galley proofs

After the copyedited manuscript has been set in type, galley proofs are returned to the author for checking. Galley proofs are copies of the typeset material without page breakdown and without footnotes, heads, and illustrations. Galleys are the first the author sees of the text material in a form similar to the way it will appear in the final book. The shock of seeing the material in type is sometimes great, as things that looked good or seemed reasonable in manuscript form suddenly look bad or ridiculous.

The galleys received by the author should already have been proofread by the editor, and any errors marked. These errors, called printer's errors and marked "PE", will be corrected along with any errors caught by the author. The galleys should also be marked by the production or copyeditor for all figure and table inserts, and the placement of all section heads and footnotes. Ideally, the copyeditor should read each galley word for word, but whether this happens depends on the budget for the book. At any rate, the author *must* read the galleys word for word. Typographical errors are bad in any book, but in a textbook they can be disastrous, as the student often is not able to distinguish sense from nonsense.

Typesetter errors that were missed by the previous proof readers should be marked by the author as "PE". These will be corrected by the typesetter without charge. Any other changes that the author makes at this point will be considered author's alterations, marked "AA", and will be billed to the author's account. Most contracts specify a limit for author's alterations of 5–10% of the cost of the typesetting. Since corrections cost much more than original typesetting, the 5–10% figure can be easily reached. Some publishers may be willing to write a figure higher than 10% into the contract, but the author should understand that any changes over the stated limit will be charged to the author and the money deducted from the author's royalty account. One problem sometimes encountered is the diligent editor who rewrites for style after the author has seen the galleys. This may not only do damage to the author's intentions, but the changes may actually get charged to the author. The author should make sure the editor marks editorial corrections as "EA" and these should then not be billed to the author.

The author must resist the temptation to rewrite the book at the galley stage. Certainly any errors caught should be corrected, whether these are printer's errors or author's errors. Extensive rewriting will not only seriously delay the production schedule, but will also cost the author money. As we have noted earlier, the time to make corrections is at the copyediting stage.

Additions of new material are not considered author's alterations, and can be added easily at galley stage. It sometimes happens that a particular piece of text could not be written earlier because the author was waiting for references or the appearance of other publications, and it makes no sense to delay production of the whole book for a few such items. A prudent author will inform the production editor that a few things may be added at galley stage, so that these can be anticipated. It is also not expensive to delete whole paragraphs of material, and the cost of these deletions will probably not be billed to the author. The main expense of correction involves changes within paragraphs. It may often be necessary to completely reset a whole paragraph just because a single word is changed. Even though the text is available in computerized form at the typesetter, there is still the necessity of finding the proper paragraph in the computer file, making the correction, running off the paragraph in a subfile, and sending it through the typesetter. Publishers are generally charged by typesetters for corrections based on the time involved in making the changes, rather than by the amount of text reset.

Most textbook authors have some familiarity with reading and correcting galleys. The formats for correcting textbook galleys are no different than those used in journals or scholarly books. Professional proofreaders have a vast array of special marks, most of which the author need not bother to learn. It is desirable for the author to learn a few key proofreading marks and then let the professionals worry about all the others. The author should follow the practice of making the correction clear to the editor, and letting the editor worry about making it clear to the typesetter.

The basic procedure is to mark the correction in the margin of the

Proofreaders' marks and symbols

Instruction	American National Standards Institute marks	
	Marginal mark	In-line mark
Delete	[delete mark]	the red book
Close up	[close-up mark]	the bo ok
Delete and close up	[delete and close-up mark]	the bøook
Restore deletion	stet	the red book
Insert in line	red	the book
Substitute in line	red	the black book
	e	tha book
Insert space in line	#	thebook
Equalize spacing	eq #	the yellow book
Lead (space) between lines	# or ld	The red book was lost
Remove leads between lines	[delete]# or [delete] ld	The red book was found
Insert hair space or thin space	hr # or thin #	100/000
Begin new paragraph	¶ or ⌊	The red book was lost. The black book was found.
Run paragraphs together	no ¶	The black book was lost. The red book was found.
Insert 1-em quad (indent)	□	The red book
Insert 2-em quad (indent)	□□ or 2	was found
Insert 3-em quad (indent)	□□□ or 3	at night.
Move to left	[	[the book
Move to right	]	the] book
Center	ctr	]the book[
Move up	⌐¬	the book
Move down	└┘	the book
Align vertically	‖ or align	The book was lost in the fog.
Align horizontally	= or straighten	The book was read.
Transpose	tr	The found book was.
		Teh book was found.
Spell out	(sp)	The (2) books came.
Push down quad (spacing material)	⊥	the book

Courtesy of the Council of Biology Editors

Proofreaders' marks and symbols

Instruction	American National Standards Institute marks	
	Marginal mark	In-line mark
Reset broken letter	x	the book
Turn right side up	9	the book
Lowercase letter	lc	the Green book
Capitalize as marked	cap	the good book
Set in small capitals	sc	am, PM
Set in italic type	ital	The Good Book
Set in roman type (Br.: upright type)	rom	the *book*
Set in boldface type	bf	The Good Book
Set in lightface type	lf	The **book**
Set in capitals and small capitals	c&sc	A Style Manual
Set in boldface italics, capitals and lowercase	bf ital c&lc	a style manual
Wrong font; reset	wf	body type
Reset as superscript (Br.: superior)	∨2	the book2
Reset as subscript (Br.: inferior)	∧2	H2S
Insert as superscript	∨4	1203
Insert as subscript	∧2	HO
Period (Br.: full stop)	⊙	Read the book
Comma	,/∧	leaves, buds and branches
Semicolon	;	Think then decide
Colon	:	Read the following
Hyphen	=/=	up and down career
Apostrophe	∨'	Lands End
Double quotes	∨"/∨"	He said book.
Single quotes	∨'/∨'	"Don't cry Fire!"
Question mark	?	Can you write
1-en dash (Br.: rule)	1/N	pages 10 15
1-em dash	1/M	The book find it
3-em dash	3/M	Ito, R. I The
Parentheses	(/)	the book a manual
Brackets (Br.: square brackets)	[/]	the book
Slant line (Br.: oblique)	/	5 ms

Example of how proofreaders marks are used

Fawns Versus Food

It is basic in animal biology that far more young are produced than necesary to carry on the species. This is true of *ants,* elephants, people and deer. The better nourished a doe is, the more fawns she produces, and the bettter chances her fawns have for survival after birth. One of the principles of deer herd management, or livestock raising, can be briefly stated if on a givn amount of food, we carry a smaller number of bred females overwinter, each one will be better fed. 10 well fed does will produce at least as many fawns as 15 half-starved ones. This has been proven beyond question. Michigan is no exception to this rule. In the upper peninsula the average rate of fawn production is 14 or 15 fawns per year from every 10 breeding does . . . and in Southern Michigan fawn production jumps up to 20 per 10 does.

—Michigan Whitetails, 1959

Fawns Versus Food

It is basic in animal biology that more young are produced than are necessary to carry on the species. This is true of ants, elephants, people, and deer. The better nourished a doe is, the more fawns she produces, and the better chances her fawns have for survival after birth. One of the principles of deer herd management, or livestock raising, can be stated briefly: If, on a given amount of food, we carry a smaller number of bred females over winter, each one will be better fed. Ten well-fed does will produce at least as many fawns as 15 half-starved ones. This has been proved beyond question.

Michigan is no exception to this rule. In the Upper Peninsula the average rate of fawn production is 14 or 15 fawns per year from every 10 breeding does . . . and in southern Michigan fawn production jumps up to 20 per 10 does.

—MICHIGAN WHITETAILS, 1959.

galley, not just in the text itself. Things marked only in the text may well be overlooked. Either the left or the right margin (or both) may be used. The only thing that should be marked in the text itself is the position of the correction. A caret is used to indicate where new material is to be added and a slash is marked through material that is to be deleted or replaced. If more than one change is to be made in a single line, the changes should be marked in the margin in order, from left to right, with a vertical or slant line separating one change from the next. If there is not enough room in one margin for all the changes, both margins can be used, as long as the order is kept clear. Every change marked in the margin *must* have a corresponding mark in the line.

Long corrections should be typewritten on slips of paper which are taped or glued to the galley, and the corresponding text which is being replaced marked in the galley. Instructions to the editor or typesetter should be circled; the standard practice is that anything circled will not be set in type. If the author is unsure about the appropriate way to make a major correction, write a note of explanation to the editor and let the editor make the actual correction on the galley.

The author should use a different color for galley corrections than either the typesetter or the editors. A sharp pen or pencil should be used, and corrections should be written clearly. If the author's handwriting is not very legible, printing should be used throughout.

The copyedited manuscript will be returned to the author along with the galleys. This permits the author to check the typesetting precisely against the manuscript and determine whether errors caught are the author's or the typesetter's. The manuscript, now called *foul manuscript,* will not be used again for typesetting, so all new marks and queries should be made on the galleys.

The *best* way to proofread galleys is to read them out loud with another person following along in the manuscript. This is a tedious, time-consuming process that many authors may want to avoid, but it is really guaranteed to find the most errors. The author should read the galleys and the helper follows along in the manuscript. The author should, of course, catch any misspellings directly while reading, and words left out or transposed will be noted by the helper. Quotations or other material taken directly from another source should be proofread carefully, and each punctuation mark read out loud. Careful attention should also be paid to the proofreading of numbers, as these often get mixed up and there is no way to detect this from the context.

Cross references to section heads should also be specifically checked. Cross references to page numbers cannot be resolved yet, and should appear in print as "see page 000". These will be flagged in the margin by the editor to ensure that they are not missed.

The corrected galleys represent a significant financial investment and should be handled carefully to be certain that they do not get lost or go astray in the mails. All corrections made on the master galleys should be transferred to the author's galleys. Then, if a set of galleys does get lost in the mail, the corrections can be transferred from the copy relatively easily.

Galleys are generally sent in batches, perhaps equivalent to chapters or sections of the book. Depending on the schedule, and how far behind the production process has fallen, there may be a rush to get these back. However, the author should insist on a reasonable length of time to read galleys. This is really the most critical step in the whole production process, and a day or two lost now is preferable to gross embarassment. Readers who find errors may blame the publisher, but more frequently blame the author. At any rate, it is the author's name that appears on the title page, not the editor's.

Dummy and page proofs

We have noted in the previous chapter that a critical stage in the production of the book is the preparation of the dummy. The dummy is really the master blueprint for the book, the instructions to the typesetter or makeup person as to how each page in the book is to appear. Every element on the page will appear in its precise location, including tables, figures, footnotes, heads, and all text. In most cases, the author will not see the dummy. The person doing the dummy should have some familiarity with the material in the book, because positioning of various elements can have a lot to do with how the book works educationally. An experienced person who has prepared dummies for a lot of similar books will generally do a much better job than a person who has not handled such books. Unfortunately, many decisions are made at the dummy stage which influence the character of the book, without the author being involved. The ideal situation would be for the author and person doing the dummy to work together, but this can rarely be arranged. For books where visual appeal is crucial, such as art books, a special effort might be made to have the author involved in the dummy process.

The dummy, along with the final typeset material and all other elements of the book, will be transmitted to the makeup person for the preparation of the final pages. The makeup person may be under the art director's supervision, or may be in the typesetter's organization, depending upon how the book is being produced and how the publisher likes to work. Once the dummy is prepared, page makeup is a relatively mechanical operation that can be done by someone not familiar with the material. However, it is a very skilled task, and one that requires a person with a high degree of compulsiveness and attention to detail.

Proofreading page proofs is generally fun, as all of the tedious work was done at the galley stage, and one is now seeing the book as it will finally appear. At this stage, it is not necessary to read the text word for word, but it should be determined that each correction made on the galleys has been properly handled. It is also necessary to check the ends and beginnings of pages to make certain that lines have not been lost during makeup. If the galleys are not returned, the author can refer to the copy of corrected galleys retained earlier (these will not, however, show any changes subsequently made by the editor).

It is important to understand that changes cannot be made at the page proof stage that will alter the length of a page. Any alterations that would affect material on subsequent pages cannot be made. If there is a major error that must be corrected by rewriting, then it is necessary to write new material that will *exactly* fit into the space occupied by the material being replaced. Problems especially come up when rewriting near the beginning of very long paragraphs. The author should remember that any changes made that are not the fault of others will be charged to the author.

Page proofs should be checked several times. During the first pass, the pages should just be examined to obtain an overall impression of the layout. If some bad figure placements are noted, they should be marked, although it may be too late to do anything about some of them.

On the second pass, all of the figure captions should be read to be certain that each caption is with its proper figure. This may be the first time the captions have been seen in typeset form, so they should be proofread word for word. Readers usually look at the figures in a book first, and any errors will be quickly noted. It is extremely embarassing (and surprisingly common) for a caption to be printed with the wrong figure. Tables will probably have been proofed with galleys, but the position of each table should be checked to be certain that it comes on the same page or a page immediately following its citation. Any problems that the editor has flagged that need author attention should be checked during this pass.

There are certain typographical conventions that editors follow. Many of these relate to problems of "bad breaks". A *widow* is a short line, with one word (or two or three little ones) occurring at the top of a page. A widow is considered bad form because it hangs alone without any surrounding words for context, and may confuse the reader. A page should also not end with the first part of a hypenated word. A section should not break at the top or bottom of the page but should be preceded or followed by at least two lines of text. Footnotes should always begin on the same page as their citations. Bad breaks are corrected by adjusting page lengths or by resetting type. Although pages are supposed to have identical numbers of lines, it is considered permissible for a page to be one line short or long. With contemporary computerized typesetters, the spacing between individual lines (the leading) can be varied slightly from page to page in order to make things fit. However, facing pages should have equal numbers of lines. In situations that are otherwise uncorrectable, a paragraph may have to be rewritten and reset. It is frequently possible for the typesetter to reset a paragraph in a tighter way so that a line can be saved, or in a looser way so that a line can be created. An even more drastic solution is to rewrite a few lines to make them shorter or longer without changing the author's meaning. Any changes needed to correct bad breaks should have been done before page proofs were made, at the dummy stage, but the author should be aware of why the text does not always appear in pages as it did in galleys. This is one reason that it is vital for the author to check pages against galleys, to be certain that everything still makes sense.

On another pass through the proofs, the running heads and folios (page numbers) should be carefully checked. Depending upon the design, the running heads may contain chapter or section titles. A common design is to have the chapter title and number on the left page and the section title and number on the right page. If a new section begins on the right page, then this section title rather than the preceding one should be used. Occasionally, the section or chapter title is too long to fit as a running head and it may have been rewritten to fit. Check the text of running heads carefully to be certain that they read well. Another duty that should be taken care of at this stage is to resolve all cross-references to page numbers. Of course, if a cross-reference is to a later page number, it cannot be resolved until that page is set. One reason to keep cross-references to page numbers to a minimum is that they are often difficult to resolve, and may be overlooked. How embarassing to have a "see page 000" appear in the final printed book! Although the editor and typesetter should never let it happen, page numbers do sometimes get mixed up. The proofs should be checked carefully to be certain that the page numbers are in sequence and that each page follows in proper order.

The author should be sent an extra copy of the page proofs for use in preparing the index. All corrections made on the master set should be transferred to the duplicate set before the master set is returned. Now comes the joy of doing the index!

The index

Last to be done, but not least in importance is the index. In a college textbook, a good index is vital, not only for the students but also for the instructors. Indexing is an art, one of the specialties of the publishing field. A busy author could hire the indexing done by a professional, but this may not be advisable even if the publisher is willing to pay, which it usually is not. Unless an unusually competent indexer familiar with the field is available, it is preferable for the index to be done by the author. Only the author knows the field and understands the nuances of the terminology well enough to decide what should be indexed. The main problems with author indexes are brevity, when done by lazy or time-pressed authors, or excessive length, when done by an author not able to be objective about the work and unable to discriminate between important and unimportant items.

Most contracts specify that the author provide the index or pay for having it done. The publisher is rightly interested in a good index for the book, and understands that a successful textbook must have a good index. However, even for an index done by the author there are expenses that the publisher should pay for. These include the cost of hiring someone to help with alphabetizing or perhaps computer time if a computer-assisted index is to be done, and these costs are a negotiable item in the contract.

There is always a rush to get the index done. It can only be done efficiently when page proofs are available, yet when page proofs are available the production schedule is of course quite advanced, and the printing date has already been contracted. The printer rightly is not interested in holding presses for completion of the index. Although indexing of pages can be done as they become available, the preparation of the final index copy cannot be done until the last pages are available. When the production schedule for the book is prepared, the date when index copy is needed should be specified, and the author should have available a two-week block of time before the due date for final preparation of the index.

The structure of the index

An index consists of a list of entries in alphabetical order and the page numbers where each entry is mentioned in the text. A simple entry consists of only a term or name and the page number or numbers. If the term is referred to on many different pages, then it is necessary to break up the entry into a series of subentries, to guide the reader to the appropriate entry. The hallmark of a good index is the presence of logical and useful subentries. Nothing is more frustrating for the reader than to look up an entry and find a string of 15 to 20 page numbers, with no guidance as to what is on each page.

Just as an entry can have a subentry, a subentry can itself have a subentry. However, it is inadvisable to use more than three levels of entry, as the user will have difficulty keeping the various levels straight.

Here are a few examples to illustrate the various levels of entry:

gall bladder, 486 (*Level one only*)

gall bladder, bile formation, 486 (*Level one and two*)

gall bladder, bile, formation in specialized cells, 486 (*Levels one, two, and three*)

A variant of the standard entry is one where a page reference is not used, but a cross-reference to another entry is included. One use of a cross-reference is when two alternate spellings may exist for a term, and it makes no sense to index under both variants. Thus, indexing all entries in a biology book under both *deoxyribonucleic acid* and *DNA* would be wasteful of space. To avoid this, when the reader looks up *deoxyribonucleic acid,* the following is found:

deoxyribonucleic acid, see *DNA*

In this case, no page entries are given under *deoxyribonucleic acid,* but all under *DNA*. Another type of cross-reference is the *see also,* entry, which indicates that additional information can be found at another place. Thus:

DNA, see also *chromosome*

Before beginning the index, the subject should be analyzed to determine the terms that are likely to encompass the key entries. The book should be studied and a list made of key terms or phrases that are likely to be indexed, such as section heads, key definitions, or major concepts. Chapter titles are such major headings that they are not likely to be useful index entries, and besides, the table of contents guides the reader to them. While the list of key entries is being made, other books in the field should be examined to see the kinds of terms they have used.

Some things are almost always indexed and are easy to include. These are names, places, and events that are referred to in relation to the subject matter of the book. However, names that are used incidentally, perhaps by way of example, should probably not be indexed. With the following question in mind, the author should try to imagine an unknowledgeable reader using the index:

Is a reader likely to look up this term in *this* book?

For instance, consider the following sentences from a chemistry textbook:

> A commonly used measure of the acidity of a solution is pH. Acids have pH values less than 7. For instance, the pH of lemon juice is 2.5.

An entry might be made to *pH*, but clearly not to *lemon juice*, because a reader would be unlikely to look up *lemon juice* in the index of this chemistry book (unless perhaps it were a text on food chemistry).

Concepts are more difficult to index than terms or names. The form of the index entry for a concept requires judgement. How will the reader likely look up the word? Take the entry:

Function of the balance sheet

in a book on accounting. This would be indexed under:

Balance sheet, function

but not under

Function, balance sheet

For concepts or ideas that span several pages, the page entry should so indicate. However, the page numbers should not span a whole section or chapter head, as this does the reader no service.

An excellent discussion of the general principles of indexing can be found in the Chicago Manual of Style (see bibliography) and in some of the references cited by that Manual.

When selecting entries, the distinction between nouns and adjectives as entry subjects should be kept in mind. Consider the following alternate entries:

DNA, chromosome

and

DNA, chromosomal

Which is preferable? It may not matter much if only one such entry is present in the whole book, but if entries like this will be indexed throughout the book, then it is important to be consistent so that after alphabetizing only one of the two will be present in the list. My preference is for the adjectival form for subentries and the noun form for main entries. Thus, I would use *chromosome* as a main entry, but *DNA, chromosomal* as a subentry. What I do is turn the subentry around in my mind and read it as *chromosomal DNA,* which sounds better to me than *chromosome DNA.*

One of the startling things about doing the index is that it generally uncovers many places where the book should obviously be corrected (misspellings, format inconsistencies, etc.). Unfortunately, it is too late to make any changes at this time, but the author should keep careful notes so that corrections can be made if a second printing is needed.

Indexing mechanics: noncomputerized

The duplicate set of page proofs that is sent to the author is used to prepare the index. All corrections that have been made on the master proofs should have been transferred to the duplicate copy. Since page proofs are sent in batches, there should be some time between batches of page proofs to do the index. In this way, the whole job need not be tackled when the last batch of page proofs arrives. **It should be emphasized that indexing not be delayed until the last batch of page proofs arrives!**

The following is a suitable way of doing a noncomputerized index if the author is able to type. It avoids retyping to prepare the final copy, and has proved to be acceptable to major publishers:

Self-adhesive labels on long rolls are used to type the initial entries. The roll is arranged so that it feeds automatically into the typewriter. The proofs are examined page by page, and each entry is typed on a separate label. Each item should be started at exactly the same place on each label, and each item should be kept on a single line. If there must be a turnover to a second line, an indentation of two or three spaces should be made. At the end of the entry, a comma is typed, followed by the page number.

It is important to index not only the text, but also all figure captions and labels, tables, footnotes, and any other elements on the page. Any *see* or *see also* entries should also be entered. If more than one way to format an entry can be used and it is difficult to decide which to use, then both should be used and the overlap rationalized later. If there is any doubt, the entry should be included, as an entry omitted now will be lost forever. To avoid boredom which might lead to the omission of important entries, no more than a few pages of indexing should be done at a sitting.

Some books use separate indexes to proper names or geographical locations. Biology books frequently use separate indexes to the Latin names of species. Such separate indexes may be valuable in large research monographs or reference books, but it is questionable whether separate indexes should be used in any college textbook. Students may overlook the separate indexes and will thus miss finding sought-for items.

There are two distinct systems of alphabetizing, letter by letter and word by word. Here is an example of the letter by letter approach:

heartache
heart block
heartbreak
heart disease
hearth
heart murmur
heartsick
heart-to-heart

The alternate, word by word, approach to the above entries is:

heart block
heart disease
heart murmur
heartache
heartbreak
hearth
heartsick
heart-to-heart

Which approach to alphabetizing to use? If there is a strong preference within a discipline, then that preference should be followed. In general, the letter by letter approach is easier for an unknowledgeable person to carry out, and if the alphabetizing is being done by a helper (strongly recommended), then the letter by letter approach should be chosen if there is no clear preference. For many books, there will be few problems of this sort, and any confusions can be handled easily by the reader.

Some special problems that have to be dealt with in the alphabetizing:

Numbers written as digits (1984, 1776, usually alphabetized as spelled out)

Initials (Smith, D.W. comes before Smith, David, even if D.W. Smith's first name is really David)

Abbreviations (DNA, FBI, alphabetized letter for letter)

Upper/lower case (case itself is generally ignored)

Prepositions, common adjectives, and articles (about, white, the, generally ignored in alphabetizing)

Practices may vary on how these things are handled, and some of the larger publishers have a house style. If the author has been given no prescription, then the Chicago Manual of Style or the McGraw-Hill Style Manual can be followed, depending upon personal preference. The key thing is consistency throughout the index.

The mechanics of preparing the index copy

Once all items have been entered on self-adhesive labels, the routine part of the job can be turned over to others. The labels are fixed individually to 3 × 5 cards which are then alphabetized. The alphabetizing of entries for a major textbook is a big job, taking a lot of space and twenty to thirty hours.

The cards are alphabetized first by main entry, and then within main entry by subentry. Within a given item, arrangement is then by page number within the book. If the alphabetizer has problems with certain entries, they can be put aside in a stack for examination by the author.

Once the cards have all been alphabetized, the troublesome job of collating the cards and rationalizing the entries begins. This *must* be done by the author, although perhaps with some help. For simple items, in which only a main entry is used, collating involves simply transferring manually all of the page numbers to the first card, and discarding the other cards. The page numbers can be written by hand, care being taken to write clearly. For the copy procedure recommended below, it is essential that all of the page numbers be kept on one line together with the entry.

For complex items, with subheads or subsubheads, some decisions may be necessary. Thus, if *DNA, chromosome* was used in one place and *DNA, chromosomal* in another, a decision must be made as to which is to be used and all page numbers transfered in sequence to that card. In some cases, an entry may have to be rewritten to make it clearer, or to permit merging page numbers more intelligently. The goal at this stage should be to reduce the total number of cards as much as possible, without doing violence to the aims of the index. Typographical errors made in typing the labels will also be evident at this stage, and should be corrected either by typing new cards or by writing on the existing cards.

Once all of the cards have been collated and all of the entries edited, it is now time to put the cards together for the final index. **The index should not be retyped!** Even if the publisher provides specific instructions for typing the index, they should not be followed. After all, the whole thing will have to be retyped again by the typesetter, and it is already in typed form on the cards. Retyping the index is an invitation to error, as well as an unnecessary expense. The manuscript for the index can now be generated by a convenient copy machine, directly from the cards, as described below.

The assistant should carefully lay the cards out in order, with each card overlapping the one above. Two columns can be laid out side by side, the number of cards in a column being determined by the output that the copy machine is capable of generating. If possible, a copy machine which reduces should be used. A page or two should be tried first, to determine how many cards can be placed in a column, and how close the two columns should be. The cards should be laid as closely together as possible, so that the maximum number of entries can be placed on a page. Once the dimensions have been defined, lines should be marked with masking tape on the work space to show the top, bottom, and sides of the layout. The publisher may have asked for the index manuscript typed in a single column. This would double the copy machine cost and would increase the amount of time necessary to make the copies. There is no reason why a two-column index manuscript cannot be used by the typesetter and my advice is to ignore a request for a single-column manuscript.

Once the cards have been laid out, each row should be taped onto the backup sheet along each side, using transparent adhesive tape. Each page should be marked with a page number and all the pages stacked upside down in order. Once all of the cards have been assembled in pages, the whole set is taken to the copy machine and a single copy made at the reduction level chosen. Index preparation is finished!

The advantage of the technique described here is that only a single typing is needed, and this typing is done by the author at the time the entries are constructed. Not only is the expense of the whole process considerably reduced, but most of the boring work can be done by relatively nonskilled individuals.

One word of advice regarding the above approach to noncomputerized indexing. The production editor should be apprised of the approach being used before the job is begun. I have used this procedure on four textbooks for the largest publisher of college textbooks in the United States, with no difficulty and without objection by several production editors. It is the best and most economical way of preparing an index without computer assistance.

The manuscript that is provided to the publisher will be sent to the typesetter for preparation of copy suitable for the printer. Ideally, the index should be proofread by the author, but this is almost never possible because of the constraints of time. The typesetter will proofread line for line, and number for number, and the production editor may also proofread. Any errors missed by them will probably never be found, except by a puzzled student. One advantage of the computer-assisted indexing described below is that typesetting errors should not occur.

Indexing mechanics: computer-assisted

A microcomputer or word processor can be used to prepare the index. The use of microcomputers is discussed in Chapter 8. The key to the successful use of a microcomputer for index preparation is the availability of appropriate software. There are many data-base management programs and file-handling programs that provide sorting capabilities, and most of these can be used for index preparation, but some of these programs are surprisingly limited. In addition to a proper sort routine, the software should also be capable of handling large files, since the final index of any textbook will likely exceed the memory capacity of the computer. A good sort program will use the computer's mass storage (floppy disk or hard disk) as virtual memory, constructing temporary files with parts of the sort until the complete job is done and then merging the temporary files into the final sorted version.

An important point in using the computer for index preparation is that the entries *must* be prepared in a completely consistent manner. There is no room for the kind of loose index construction that might be satisfactory when a human is doing the sorting and collating. Any inconsistencies will show themselves in improperly sorted entries, which will be embarassing if they end up in the final book.

How a computer sort works

It is important to understand that the computer sorts items on the basis of the numerical codes that are used to represent the letters and symbols. The code used by all computers is ASCII, which is reasonably logically constructed but has some peculiarities and limitations. The most serious problem is that the ASCII code for an upper case letter is different than the code for a lower case letter. Upper case letters have code values 32 units lower than lower case letters. Thus, "Gambrinus" will not sort between "gall stone" and "gamma radiation", but somewhere completely differently, unless the sort program can be told to ignore case (surprisingly, many commercially available sort programs cannot ignore case).

The sorting begins at the leftmost letter of an entry and moves through the entry letter by letter. A space is treated as a separate letter, so that "heart block" will not sort between "heartache" and "heartbreak", but before both of these (since the ASCII code for a space has a lower value than the code for any letter). An initial with a period after it will sort differently than an initial without a period, because of the code for the period (decimal point).

Another major problem with computer sorts arises if italicized words are in the index. This is a special problem in the field of biology, but does arise to some extent even in other fields. If coding for italics is to be included in the entry at the time it is typed, then this coding will force these words to be sorted separately from nonitalicized words. One procedure is to not add codes for italics at the time the entries are typed, but if a large number of italicized entries are being used, this presents complications at a later stage.

It is easy to appreciate that a large number of exceptions which a human sorter can easily take care of must be specifically planned for if a computer sort is to be used.

Typing the entries

The entries must be typed in a consistent manner, and every space counts. In addition, the second and third level entries must be coded in some way, so that they can be handled subsequently. A reasonably easy way of doing the actual typing is to use a slash as a separator between first, second, and third levels. Thus:

Darwin/Voyage of the Beagle/341
Darwin, Charles/325
Darwin/Origin of the Species/375

would be three separate entries that would ultimately be collated under Darwin and would appear in the final index as:

Darwin, Charles, 325
Darwin:
  Origin of the Species, 375
  Voyage of the Beagle, 341

When the computer sort is carried out, these three entries would sort in alphabetical order with the slash separators still in place. A subsequent program would then have to be used to examine each sorted entry for the presence of the second level entries and to put the page numbers in proper location. Note that in the example, one of the entries does not have a second level, whereas the other two do. In the final index, the entries without the second level would occur first, followed by those entries with second levels, in alphabetical order of the second level entry.

Once the entries have been sorted, the complete index should be printed out and carefully edited for accuracy, consistency, and format. Entries should be combined and deleted, revised for proper syntax, etc., exactly as has been recommended for the noncomputerized indexing. The revised version is then edited again and a manuscript of this final version prepared for submission to the publisher.

Although software does exist for handling all of the problems referred to here, such software is not widely available. An author who is interested in using a computer for preparing the complete index for a book should contact the American Society of Indexers (address in Literary Market Place) for advice. This organization not only has a number of members who are experienced in doing computerized indexes, but has members or affiliates who offer software that can be used by authors. Because of the continual state of flux of the computer field, it is difficult to provide any detailed recommendations for index preparation.

The finished book: corrections and revisions

After the index copy has been sent in, the author's work is finished. A month or two later, bound books will be available and the production editor should send an advance copy. Finally, everyone will see what all the fuss has been about!

For some authors, seeing the bound book is anticlimactic (mothers frequently feel a similar letdown when their first baby is finally born). There has been so much attention to boring detail over the past few months of production that the sight of the completed product is deflating. However, for most people, the arrival of the first bound book is an occasion for celebration. There it is, finally finished!

Now that the book is completed the author can put it on the desk and go on to other things, patiently awaiting the arrival of the first royalty check. Right? Wrong! There is still work to do, and it involves checking the book for typographical errors.

The successful textbook will have more than one printing, and minor typographical errors should be corrected for the second and subsequent printings. Correcting typographical errors is not only a service to the reader, but will ease the burden when it comes time to revise the book for a second edition three or four years hence (the successful author will not be allowed to rest on any laurels, no matter how hardearned). It is likely that an amazing number of typographical errors have passed successfully through all the proofreading stages and are there in cold type staring out at the reader. How to find them?

Most textbook authors, busy with a multitude of tasks, are not inclined to sit down and read through the book word for word again, and I certainly don't recommend this. My favored approach is to pay graduate or undergraduate students in my department $1 an error for all errors found. Each error is paid for only once and the first student to find a given error gets the $1. I keep a master copy of the book in my office in which I record all errors as they are found, and use this copy when it is reprint time. I have found that students enjoy finding typographical errors, and some of them are quite good at catching blatant errors as well as the odd misspelling in a remote footnote. I serve as the final judge as to whether the found item really is an error, and pay up on the spot. The expense of this procedure is usually not great.

Most publishers will have a standard procedure for making corrections for second printings. The reprint editor will contact the author well in advance of when the printing is to take place and will request a list of corrections. The corrections that can be made are precisely the same kinds of corrections that can be made at page proof stage. Thus, corrections that involve a change in paragraph length or pagination cannot be done. Only real tyographical errors can be corrected at this stage, and major rewriting will have to await the second edition. Hopefully, the list will not be longer than one or two pages.

In addition to simple typographical corrections, the author should also begin to compile a list of more major changes that should be made for the second edition. As the book is examined, obvious problems that need changing will become apparent. The author should also receive correspondence from colleagues and others using the book who have found problems or mistakes (major and minor). A file should be kept with all of these letters and notes. When the author talks with colleagues about the book, they should be given an enthusiastic invitation to send in errors or problems. In some cases, the colleague may be wrong and the author correct, but even the wrong suggestions will be valuable, as they will force the author to think about how well the job has been done.

If the book has been written properly, and the publisher has found the right market, the book should be a big seller, and the author will be well on the way to a profitable part-time career as a textbook author. The book should be viewed as a long-term investment, almost like an annuity.

How the author can help to keep costs down

Book production is a complex process, not amenable to assembly-line tactics. Sometimes minor things that are done wrong along the way can have major effects on the cost of book production. Since the final price that the book is sold for will be determined to a major extent by production costs, it is important to do everything to hold costs down. Hopefully, the publisher's production staff will be attuned to cost containment, since the production staff's activities have the most impact on the final cost of the book. But there are things that the author can do which will help to control costs. It should be remembered that costs saved here may end up as extra spending money in the pockets of students who would otherwise have to spend more for the book.

- Make all your major changes before the manuscript is sent to the publisher.
- Make all changes missed in item 1 on the copy edited manuscript, before typesetting.
- Avoid as much as possible major changes at galley proof stage. Re-

serve this stage just for correction of typographical errors and obvious errors of fact.

- Do not redesign illustrations once they have been drawn. Artist's errors should be corrected by the artist gratis, but any major changes from the copy you have supplied will be charged extra.
- Do not make any changes at page proof stage except for corrections of obvious errors. No changes at this stage should alter the spacing of text or the length of paragraphs.
- Avoid expensive air freight charges by reading copy and proofs quickly upon receipt, and returning by less expensive air mail.
- Delays due to slowness on your part may cost money if publisher's employees or suppliers are sitting idle waiting for material from you. You should make every effort to maintain the schedule that has been set.

In addition to helping control costs of the final book, attention to the above list will also put money in your own pocket, because, as we have noted, the author is charged for alterations above a certain percentage, and postage expenses from the author's end are generally borne by the author rather than the publisher.

An author's checklist of book production

Here is a brief list of key steps in the book production process that impact on the author. Cognizance of these points should help to make the partnership between author and publisher a smooth and uncontroversial one.

1 Complete the manuscript by the date agreed to with the acquisitions editor, as there is an optimum time for the book to appear, and even a small delay in the schedule may result in a major delay in the book production process.

2 Once the manuscript is submitted, be certain that it is received by the acquisitions editor, and obtain confirmation in writing of receipt of an acceptable manuscript. Written confirmation has legal implications if an advance or bonus for completion before a given date has been specified in the contract.

3 As soon as the book is launched into production, make contact with the production editor and establish a firm, pleasant working relationship. Discuss especially the production schedule and determine that the tasks you will be required to do will not conflict with travel plans or major duties.

4 Go over the copyedited manuscript carefully and make all major changes at this stage. Check to ascertain that the copyeditor has not done violence with the facts in the process of rewriting the syntax. Do not allow the copyeditor's changes to annoy you, even if wrong. Be firm and forceful, but reasonable.
5 Answer all queries on the manuscript in writing.
6 Read the galleys directly against the manuscript. The best procedure is to have a helper follow the manuscript while you read aloud.
7 Remember that you will be charged against royalties for excess alterations of material already set in type (the percentage should be specified in the contract).
8 Check the artist's sketches carefully against your copy, making certain that positioning and labelling are correct. It is much cheaper to make changes on the artist's sketches than on the final drawings.
9 Changes made at the page proof stage are very expensive and should be made only when absolutely necessary. No change can be made in page proofs that will alter the length of paragraphs or other elements. If a change in a paragraph must be made, the text will have to be rewritten so that the spacing remains the same.
10 In the later stages of the production process, when difficulties are often experienced in keeping to the production schedule, the author may be handling many points by telephone with the production editor. It is essential that the substance of all telephone conversations be confirmed in writing, since agreed-upon items are easily forgotten.
11 Begin to do the index as soon as page proofs start. There is usually a few days' lag between correcting and returning one batch of page proofs until the next batch comes. Use that time to do the initial index entries. Do not wait until all the page proofs are available to begin the index!
12 Once the final book is in your hands, check it carefully for typographical errors that can be corrected in the second printing, and begin to assemble a file of readers' comments and any changes you deem essential for the second edition.

Good luck on your sales!

8

Computers and word processors

Although many books are still written using conventional typewriters, for most major textbook projects computers or word processors are used. The advantages of the computer are great, and the cost modest (compared to the other costs involved in producing a textbook). When beginning a completely new textbook, the decision to use a computer is especially easy to make. I have written four complete books using a microcomputer, and would not do it any other way. The procedures for book production described in other chapters have been mostly written without specific mention of computers and word processors. It is the purpose of the present chapter to present the general procedures used and the precautions needed when using a microcomputer or word processor for writing a book.

At one time, there was a clear distinction between a microcomputer and a word processor, but this distinction has virtually disappeared, and for the purposes of this discussion the two will be considered equivalent. Many textbook writers will already have access to a word processor or microcomputer. If one is not available, the acquisition of one just for this textbook project should be considered. The cost of acquiring a microcomputer for a major textbook project should be borne at least partially by the publisher. Most publishers are willing to provide an advance on royalties to cover the purchase cost, and if the project is large enough (say, a major freshman/sophomore-level textbook), they may be willing to pay for the purchase outright. There are so many advantages to the publisher of having the author do the book on a word processor that the author should find little resistance to a request for purchase. For a book with low anticipated sales, where the book budget cannot bear the cost, the author should consider making the investment from personal funds. Any purchase of a microcomputer for professional purposes is probably deductible from the income tax (but not all at once; the purchase will have to be capitalized and depreciated over a number of years). If an outright purchase seems inadvisable, then the author should at least consider leasing a system for the period of the writing project.

Microcomputer hardware

I do not propose to make specific recommendations on microcomputer hardware. There are many microcomputers on the market that will serve the textbook author well. The author would be well advised to purchase equipment from a local retail dealer who is able and willing to provide support and service. Whatever is purchased, it can be assumed that it will become obsolete within the next five years because of new developments. Listed below are the hardware features necessary for a microcomputer that is to be used on a major textbook project.

- Central processing unit
- Memory
- Mass storage
- Video display
- Keyboard
- Communications capability; modem (possibly optional)
- Letter-quality printer

I discuss each of these briefly below.

Central processing unit

The central processing unit (CPU) is the board or chip in the computer which the handles calculations and interprets program instructions. The average user will neither know nor care where the CPU resides. The main concern about the CPU arises from the fact that word-processing software is CPU-dependent, so that if a particular program is desired, then it is necessary to have a microcomputer with a CPU suitable for that program. The textbook writer is strongly advised to decide first on the software that will be used (see later), and then select a hardware system containing a CPU that will run that software.

The CPU can be 8 bit, 16 bit or 32 bit. This refers to the number of binary digits that are operated on at one time. Calculations can be done noticeably faster with a 16-bit CPU than with an 8-bit chip, but for most word-processing functions, the difference in speed is insignificant. Another advantage of a 16-bit CPU is that it can address a larger memory bank directly, but 8-bit machines have developed methods for handling larger memories. A single character (e.g. letter, number) can be coded for with only 7 bits, so that there is no problem in processing text using an 8-bit machine.

Memory

For word-processing functions, a large amount of memory is desirable. The computer should have sufficient memory so that a reasonable amount of text can be in memory at once. Even though a mass storage device (see below) may serve as a virtual memory, there are some major advantages to a large amount of memory. At the very least, a computer with 64K of memory is needed. Since some of that memory will be occupied by the computer operating system and the software, considerably less than 64K will be available for text. For a major project, 256K of memory is advisable. One point of caution: make sure the word-processing software used can access all of the available memory.

Mass storage

A mass storage system is a device on which large amounts of computer data can be written for permanent or semipermanent storage. The most commonly available mass storage device is the floppy disk, but tape and hard disk systems are also used. Good word processing software will use the mass storage device as equivalent to memory (called "virtual memory"), and this makes possible the editing of larger amounts of text in a single file than would otherwise be possible. Floppy disks have various amounts of storage possible, from 150,000 to 500,000 or even a million characters. Hard disks can store much more data. A hard disk would be capable of storing the complete text for several books, plus all the necessary software. The advantage of a floppy disk over a hard disk is that the former can easily be taken out of the machine and transported somewhere else, for use on another machine. It takes about four floppy disks to store the text of a book of average length. One disadvantage of a hard disk, in addition to higher cost, is that backup is difficult. Although hard disks are reputed to be very reliable, they do crash from time to time, with the potential for destroying all of the work on a lengthy book.

An important thing to consider in selecting the mass storage device is that the speed of access to the device by the computer should be high. During word-processing applications there will be many accesses to the device, and a slow-acting device will considerably slow down many aspects of the text editing function. If a hard disk is used, then provision for backup on some other mass storage device should be available.

Keyboard

Although the main number and letter keys are standard on all computers and typewriters, the keys for extra characters and the function keys are definitely nonstandard. Although a decision to purchase should not be made strictly on the basis of the keyboard, a user should try the keyboard to be certain that there are no major difficulties. Of course, the keyboard must be able to generate the full set of characters, including upper and lower case, numbers, and punctuation. Computers use a coding system called ASCII (American Standard Code for Information Interchange), and the full ASCII character set consists of 128 characters. The first 32 characters of the ASCII set are unprintable control characters (things like "carriage return"), and most of these are generated by typing other keys together with the "control" key. Most of the rest of the ASCII characters are printable, and are generated either by pressing a key directly, or by pressing a key together with the "shift" key. Although some of the ASCII characters are little used in straight text, they find wide use in computerized typesetters and will be very useful if the computer is to be interfaced with a typesetter (strongly recommended, see later). Thus, the full ASCII character set should be available on the keyboard.

One nonprintable ASCII character is DELETE. Much word-processing software uses the DELETE key to permit back up and erasure. Note that the DELETE key is used in a different way from the BACKSPACE key, because with the latter the cursor backs up without erasing.

Other keyboard features, such as a separate number key pad, or tab and cursor positioning keys, are useful but not essential. It should be understood that all that any key does is generate a specific code. How this code is interpreted is a function of the operating system and the software of the computer. Thus, keys that appear on the keyboard with such designations as LINE FEED or BACK TAB may do something quite different from that, or nothing, if the software is not prepared for them.

Video display

The video display should provide for at least 80 characters horizontally, and 24 lines vertically. Some dedicated word processors have full-page display capability, which is especially valuable when looking for page breaks on long pieces of text, but is not especially needed for book writing. Flat-panel displays function similarly to video displays, but do not require as much space.

The user should be happy with the legibility of the character set on the video display, and with the color and contrast of the characters. Most human engineers apparently feel that green characters on a black background are the most restful to the eyes.

The video display unit and accompanying software should support reduced or reversed video (or preferably, both). With reduced video, the intensity of the characters is about one-half normal, and with reversed video, the characters appear black on a light background. For word-processing functions, reduced video is commonly used to separate instructions from text, or to indicate blocks of text which have been marked. For a major book project, this is very desirable (and software should be chosen which will support this function).

Communications capability

One method for interfacing a microcomputer with a typesetter is by the transfer of the final text over a telephone line connected to the typesetter. To do this, a communications capability is needed. The simplest device is an acoustic coupler into which the telephone headset is placed, which is connected to the computer by a standard serial interface device. A preferable arrangement is a modem unit which either fits into one of the expansion slots inside the computer or is a stand-alone device which is accessed through an interface port on the back of the computer. Most modem units permit not only direct connection to the phone line, but also automatic dial and (with proper software) file transfer.

The rate at which data are transferred over the telephone is a function of the modem. The slowest transmission rate suitable for serious work is 300 baud (300 bits per second, equivalent to around 30 characters per second), but 1200 baud is better. The receiving device must be able to accept characters at the same rate they are being sent. A key part of any telephone transmission system is the software which operates it. A special terminal program will be needed to permit operation of the communications modem.

One of the major advantages of transmission of text via the telephone is that it permits communication between computers that would otherwise be incompatible, so long as both use the ASCII code. Incompatibility of disk formats or CPUs is of no consequence provided each computer is capable of sending and receiving ASCII. In fact, it is possible to convert text from one format to another by transferring over the telephone.

Communication between computers does not absolutely require a telephone. If the two computers are in the same room, they can be connected directly, without any intervening telephone line. This may be useful if a large amount of text is to be sent, since it does not require tying up a telephone line.

Letter-quality printer

The printer is a major part of the purchase cost on any system, and it is frequently tempting to save money by buying a cheaper printer. Don't! A serious textbook writer needs a serious printer. The term used here is "letter quality", which means that the print looks like it was done on a conventional typewriter. Get a printer that not only generates good looking type, but one that is also sturdy and reliable. The printer will get a lot of use before the book is finished, and should be able to stand up to heavy use. (My printer, a Diablo 1640, has printed four complete books, including several drafts of each, without needing a service call.) For preliminary drafts that only the author will read, a cheap printer could be used, but then a letter quality printer will have to be located that can be tied up for a long period of time for the main printing.

Appropriate software is needed to make a printer do everything it is capable of doing. For example, underlining, superscripts and subscripts require that the printer do special things, such as backspace and half-line feed up and down. Not only must the printer be able to do these things but the software must support them. It is strongly advised that the author see the printer work *with* the computer *with* the software before any purchase is made. A printer should be selected for which good service is available in the area.

If the printer is to be used with a microcomputer which uses bit-mapped graphics to display characters (such as the Apple Macintosh and Lisa), an additional consideration must be given. The characters seen on the screen of such computers do not represent simple ASCII codes that have been generated by a character generator inside the computer, but are actually constructed pixel by pixel on the screen using the computer's own software. This permits the generation on the screen of a variety of fonts (bold face, italics, Gothic, etc.), but the only way these characters can be printed is by use of a dot-matrix printer that transfers every pixel from the screen to the paper. Frankly, even the best dot-matrix printer does not exhibit the print quality of a letter-quality printer, so that the text output from a bit-mapped graphics computer leaves much to be desired. Translation software may be available that will convert the bit-mapped graphics into normal ASCII for output to a printer (although then, of course, the font seen on the screen will not be seen on the paper), but it is essential that careful tests be run to be certain that the letter-quality printer and the bit-mapped graphics computer are compatible.

Software for textbook writing

Although the hardware costs the most money, it is the software that makes the system work. Most computer consultants advise that the user select the appropriate software first, then buy the hardware capable of running it appropriately. Although there is a lot of microcomputer software available, useful for everything from income tax preparation to games, my discussion here deals only with the software needed for text processing. Word processing software is available for virtually every microcomputer on the market. Some is useful only for such simple matters as writing letters. However, most microcomputers will have available software capable of supporting the writing of a whole book. Because software developments are fast, and software companies rise and fall with the quality of their management, I shall not make software recommendations by name, but shall describe the desirable characteristics in sufficient detail so that a purchaser can make an intelligent decision from the available products.

Here are some features that should be present in any word-processing software used for writing books:

1 Uses the standard operating system of the microcomputer.
2 Creates standard files in mass storage (usually floppy disk) that can be interpreted by other programs.
3 The size of the file that can be edited should not be restricted by the memory size of the computer, the mass storage device being used as a sort of virtual memory.
4 It should be possible to merge smaller files into a larger file, so that the text can be written and edited in small blocks that can be merged for the final copy.
5 It should be also possible to mark blocks of text within a given file so that they can be moved around and written to separate mass storage files.
6 The files created by the word-processing software should interface to spelling programs.
7 The files written should be transferable over a telephone line to another computer (useful for typesetting).
8 The floppy disk format used should be one which is widely available, so that the disks can be read by other systems.

Word-processing software contains two separate modules called the *text editor* and the *text formatter.* The text editor is used to create and edit the computer file containing the characters written. The text formatter is used to display the text in properly formatted form on the printer. A useful word processor should exhibit the text on the terminal in a form similar to that in which it will appear on the printer. Many good word-processor packages do this, showing on the screen the text with proper margins and page breaks, although not with proper underlines and subscripts or superscripts. The word processing software should handle these latter formatting features of the printer.

Here is a list of some of the editing features found in good word-processing software:

1 Text should scroll up and down the screen, either line by line or screen by screen.
2 The cursor should be able to move forward or back through the text either by character, word, or screen.
3 It should be possible to delete individual characters, words, or whole lines, and it should be possible to insert new text at any place within the old text.

4 It should be possible to set margins, line spacing, and tabs.
5 Text should be formatted automatically for margins, left and right justification, and paragraph indents as it is typed.
6 It should be possible to center a line of text within the selected margins.
7 It should be possible to mark a block of text for deletion, duplication at another location, movement to another location, or storage as a separate file in mass storage. It should also be possible to bring a block of text in from the mass storage device.
8 The word processor should have search and replace functions. It should be possible to search a file for a particular string of characters, and to replace a search string automatically with another string. It should be possible to do search and replace globally (throughout the whole file without operator intervention).

One useful feature of most word-processing software is that the length of a file, in characters or words, can be readily determined. Since most author's contracts specify manuscript length, this makes it possible for the author to adjust the writing to the desired length.

In addition to the editing features listed above, the word-processing software should support a number of features for displaying the text at the printer:

1 Pages should be numbered automatically.
2 It should be possible to automatically include a header or footer on each page, containing a section or chapter title.
3 It should be possible to set the length of the page to any value, and to have printing begin on each new page automatically.
4 It should be possible to underline text automatically, and to print subscripts and superscripts. These latter features require not only the appropriate word-processor software, but a compatible printer capable of half line spacing up and down.

With the above software and hardware, it should be possible for the textbook author to write a book of virtually any length and generate a high-quality manuscript. The flexibility of such a system makes it possible to create error-free text in the shortest possible time.

Some advice on using word processing

There will be a significant amount of time in the early stages of using a word-processing system in which difficulties will arise. The user must be patient and avoid discouragement until this stage has passed. Once the system has been mastered, things will seem much easier and in the long run, the advantages of using a word processor will be obvious.

Naming of computer files is always a little bit of a problem, because most systems have a restricted number of characters that can be used in the name. A naming system is needed which reveals something about what is in the file, and which permits updating as more versions of the book are created. In computer file names, every character is recognized, and if two names differ by even a single character they will be considered different files. Most computer file names must begin with a letter, although numbers and special characters such as dashes can generally be used after this. A suitable naming system for a book is to use names such as CH1-1, CH1-2, CH1-3 for the subfiles within a chapter. When CH1-1 is revised, the same name can be used for the revised version, or another name can be given, such as CHP1-1, if it is desired to keep the several versions separate. After the final chapter is written and all of the files have been merged, a name such as CH1 can be used. For a book with numbered section heads, each section head could be a separate file and the section numbers could be used as file names, such as SC9-3, which refers to the third section in chapter 9. When the chapter is complete, all the sections are merged together and the file named CH9.

Always keep two copies of each file! Although computer accuracy is high, mistakes do occur, and computers sometimes die in the middle of a task. Whenever writing and editing of a new section is complete, file it immediately in mass storage and make a backup with whatever backup system is available (probably another disk). If you are using a removable disk system for mass storage, store the backup disk in another location, to protect against loss due to fire or theft. Remember that a lot of time went into writing this material, and time is equivalent to money. Finally, before turning the computer off, **print** the newly written file so that a hard copy is available. Always maintain two hard copies of everything written, each stored in a different location, so that lost files can be recreated, if necessary, by retyping.

Keep files that you are editing short, no more than ten pages per file. Short files are easier to edit than long files, and can be always be merged for the final version. When the hard copy is printed, immediately mark the file name used at the top of the first page, so that this file can be found again in mass storage.

When the reviewer's comments have been received and revision is about to begin, make a new set of file names for the revised version, and store these on a new set of disks. Keep at least one of the original versions as another backup, until the revision is completed and a backup of the revised version is available.

Other useful software

In addition to the word-processing software discussed above, some other software is useful in a major textbook project. It is desirable that the ancillary software discussed here be compatible with the word-processing software. Most word processing software companies provide programs such as those discussed here, and if all programs are bought from a single company, compatibility is assured.

Spelling programs

One of the most useful programs for any textbook project is a spelling checker. This is a program that will check the text for misspelled words. The spelling program will go through the textfiles and check each word against a large dictionary of commonly used words. Any word not found in the dictionary will be flagged. There is a provision in most spelling programs for a supplemental dictionary of words not found in the major dictionary. Since any academic discipline has a large number of specialized words that will not be present in the standard dictionary, when the spelling program is first used, such a dictionary will have to be prepared. The procedure for doing this is to go through the words that the spelling program has flagged as misspelled. In the absence of a supplemental dictionary, all of the technical terms will be flagged, and these can then be marked for inclusion in the supplemental dictionary. The spelling program should permit construction of this supplemental dictionary from the flagged terms. After a few chapters have been checked, most of the technical terms will have been caught and then only truly misspelled words will be flagged.

A spelling program is of special value if the files are to be transferred to a typesetter without rekeying, as recommended below. In essence, all of the proofreading of the text will have been done before any serious typesetting has been carried out.

Spelling programs frequently have accessory capabilities that may be useful. Most such programs will count the number of words in a document (although the word-processing software may also do this). Some spelling programs will also do textual analysis, such as counting the frequency of occurrence of certain words, or flagging homonyms, and a few may also do grammatical analyses such as looking for passive voice, but the utility of most of these features for the college textbook writer is questionable. What the writer does need is a spelling program that operates quickly on large quantities of text, has a large standard dictionary, and has the capability of generating a large supplemental dictionary.

Sorting programs

A very useful program for the preparation of the bibliography and index is a program which sorts. Sort programs can either be stand-alone programs that operate on files created by a word processing program, or data base management programs that permit creation of files as well as sorting. If the latter kind of program is used, it should be ascertained before the program is purchased that the files created by the data base management program can be edited by the word-processing program. This compatibility is essential for any serious use.

Sort programs work by rearranging records so that they conform to a desired order, generally alphabetic or numeric. A "record", as far as the sort program is concerned, is one of a number of identically formatted blocks of text. A block of text could be a single line, or a group of lines that are terminated in some way by a character recognizable by the program (such as a carriage return). Within each record, sorting will operate on one or more fields. A "field" is a block of text within the record that has a defined position so that it can be found by the sort program. The position of a field within a record can be marked either by its absolute position (that is, so many characters from the beginning of the record) or by use of special field delimiters around the block. The simplest way of thinking about records and fields is in relation to a bibliography. The complete citation of a single article or book is equivalent to a record, and fields within the record would be such things as "authors", "date", "title", and "journal citation". Sorting could operate on any of these fields, or on two or more fields together. The commonest way of sorting bibliography records is by the two fields "author" and "date", the "author" field being used for the primary sort and the "date" field being used for the secondary sort. After the sorting program completes its work, the bibliography is arranged alphabetically by author, and by date for all entries for the same author.

Clearly a sort program is very useful for a textbook project, since it permits entry of citations in the order in which they are used in the text, and then rearrangement in the form needed for the bibliography. It is essential, however, that all of the records be arranged in exactly the same way, because the sort program has no way of checking for mistakes. Thus, if a given field is missing for a record (for instance, a citation without a date), then some provision for this missing piece of information has to be made when this record is first typed. Another problem with citations is that journal articles and books have different forms of citation, and an arrangement must be adopted that keeps these two kinds of records compatible.

A major problem with sort programs is that they alphabetize character by character, starting at the beginning of the field. It is the ASCII code of the character that is being checked, so that upper case and lower case characters are treated differently. Some sort programs will have a provision for ignoring case, and it may be desirable to use this feature. Another problem with sorting occurs in multi-author citations. In most cases, the author field will include all of the authors, and the sort program will therefore move across the field, comparing one character after another of the field to that of all the other records. If all citations are arranged in the same way, with the last name of each author preceding the first name, then no problem will arise, but if some citations are arranged one way and some another, a mixed sort will occur. Thus, it is essential when setting up the format for a series of records that all fields be properly arranged.

Once the initial sorting is finished, it is necessary to examine the sorted records with the word processor to be certain that the records are properly arranged. Sorting is an excellent way of catching certain kinds of mistakes, which can then be corrected with the text editor.

It is important that any sort program used be able to sort and combine multiple files. In a textbook project, for either the index or the bibliography, a large number of records will be sorted, and the initial entry of these records will likely be in a number of separate files, perhaps equivalent to chapters. Once the records have been properly entered and stored, the sort program can then merge these multiple files into one large file for final editing and printing.

In addition to its use with bibliographies, a sort program is extremely useful for the preparation of the index. The preparation of an index was discussed in Chapter 7.

From word processor to typesetter

When text is typed on a word processor or microcomputer, it is converted into a form which can be interpreted by other computers. Since typesetting systems are all computer-controlled, they are able to process text and put it into typeset form, so long as proper instructions have been provided. In the terminology of the trade, any text written at a computer exists in "machine-readable" form. In the act of creation, the keystrokes are "captured" and can thus be interpreted by other computers. Given proper interfacing, any microcomputer or word processor can be considered the text entry portion of a typesetter.

Given the above situation, it seems foolish for the carefully prepared text, all proofread and formatted, to be typed all over again by a typesetting keyboard operator. Not only is this costly and time-consuming, but it is error-prone. The goal, then, is to have the machine-readable text drive the typesetter. This saves considerable time and money, and has the potential of creating a better book.

Unfortunately, there are a number of new problems that arise when interfacing a microcomputer to a typesetter. Although the procedure is to be strongly recommended on any new textbook project, to do this properly requires considerable advanced planning and preparation. Although a number of textbooks have actually been prepared by authors in this way, success requires a sympathetic publisher and a detailed understanding of the whole computerized typesetting process.

To understand the problems involved in such a project, let us first consider how typeset material differs from typewritten material. What things do typesetters do that typewriters don't?

- The fonts (forms and sizes of characters) are much more variable on the typesetter. A wide variety of typefaces is available, and within each typeface there are various categories, such as **bold** and *italic.*
- The typesetter has available a wide variety of special characters such as foreign alphabets (α, Δ, Θ, λ, é, Ö) and symbols (©, °, ™, ℞, ⬚, £, ♂, ♀, △, ★).
- The typesetter is able to produce a wide variety of mathematical symbols and constructions ($\times$, $\pm$, $<$, ≂, $\supset$, ∻, ∞, ⇦) and to display formulas in many intricate ways.
- The typesetter is able to make subscripts and superscripts, with the scripted characters being of smaller size (A^1, B^3, T^{growth}).
- The typesetter is able to carry out table alignment and to create both horizontal and vertical rules.

- The typesetter is able to carry out proportional spacing, ensuring that characters such as "i" and "l" take up less space on the page than characters such as "m" and "t". On the typewriter, each character occupies exactly the same amount of space.
- The typesetter is able to justify text (so that the right margin is aligned) by putting a variable amount of space between words, or even between letters.
- The typesetter is able to hyphenate automatically.
- The typesetter is able to make changes in the sizes of characters for headings, footnotes, etc., and to insert extra vertical space where needed above and below headings.

Some sophisticated typesetters can do even more things than those listed above, but these are the basic capabilities that all typesetters have. These characteristics, of course, are what make typeset material look so much nicer than typewritten material, and why most books are typeset. There is a considerable improvement in readability when text is converted from typewritten to typeset form. In addition, the material is considerably compressed in the area it occupies on the page (usually by about 25%).

Now comes the rub! How do we ensure that the typesetter does what we want it to do? Obviously, all of these special characters, changes in font, subscripts and superscripts, display of mathematics, etc., that the typesetter can do will not be done unless the typesetter is *instructed* to do them. And here is where the author enters into the picture. Capturing the keystrokes in machine-readable form is one thing, but inserting typesetting instructions is quite a different thing. Although the word-processing files are readable by the typesetter, they cannot be typeset without inserting typesetting commands. These codes are either inserted by the typesetter, or by the author at the time of typing.

In the terminology of the typesetting trade, "markup" refers to the insertion of typesetting commands into the text. In the days of metal type, markup was done on the manuscript itself, and involved writing instructions that the keyboard operator would interpret. In the computer age, markup involves the insertion of symbols into the file that will be interpreted by the typesetting computer as commands or instructions rather than text. A system is necessary that ensures that certain characters are interpreted not as text but as commands.

To understand this a little more, a brief discussion of the components of a modern typesetter must be given. The typesetter consists of two main parts, the phototypesetter itself, which generates the characters and displays them on the photographic paper or film, and the typesetting computer, sometimes called the "front end", which takes the text and the embedded commands and converts them into instructions that can control the phototypesetter. In sophisticated front-end systems, very elaborate control over the typesetter is possible, and output can be either galley proofs or pages made up exactly as they should be to go to the printer. In a complicated textbook, with many illustrations and problems of page makeup, it is difficult to output pages directly from the typesetter, and galleys are generally preferred. This makes the author's job much easier, since only a minimum of markup commands must be inserted.

The typesetter front end is a computer which itself has all of the word-processing capabilities. It has the capability of editing text, doing complex search and replace operations, and converting text marked in one way to text marked in a way that it can handle. Thus, a good front end will place few restrictions on the markup that the author must incorporate. The ideal arrangement is for the author, production editor, and typesetter to agree on a simple markup system which the author can use, and then have the front end convert the text into a form suitable for the typesetter. In this way, the author can generate machine-readable text that is typesettable, without having to become involved in learning the arcane aspects of typesetting itself. A markup system of this type is sometimes called a "generic code", because it is not tied to any particular brand of typesetter. Several generic coding systems have been developed, and standardization of generic coding will simplify the conversion process from microcomputer to typesetter. However, any typesetting system sophisticated enough to be used on a major textbook should be able to interpret an author-devised generic code. A generic coding system advanced by the Association of American Publishers is included in the Appendix.

At this point, it is important to understand a key point. The formatting that the text has been given in the word processor is not directly interpretable by the front end. Thus, if a line is to be centered, or a word underlined, or a paragraph indent or tab made, these things will be handled quite differently by the typesetter. Right-hand justification, hyphenation, and other spacing instructions that might have been put into the text will be useless and will only confuse the front end. When the text is formatted in the typesetter, quite different spacing will be used, line breaks will come at completely different places, and things such as underlining will not be used (underlines are usually converted into italics). As much as possible, all of the word-processor formatting commands should be removed from the text that will be sent to the typesetter. If they are not inserted in the first place it is even better. However, media conversion services exist that will, for a price, translate word-processing formats into typesetting formats, making possible direct conversion, provided that only simplified formatting is required (for instance, no special characters).

In thinking about how the text will be handled, it is convenient to distinguish between two kinds of commands, those that must be inserted into the body of the text itself, and those that are outside the body of the text and which control such things as section heads and footnotes. The text should be thought of as a series of paragraphs, and each paragraph as a long string of characters. When a paragraph is normally typed, a specified paragraph indent is inserted, and at the end of the paragraph a special command (probably an extra carriage return and line feed is used). For the typesetter, the paragraph indent is not needed, as the typesetter will incorporate its own indent which will be different. The end-of-paragraph marker will be needed, and this will have to be one of the typesetting commands that will be put into the text when it is being typed. Since lots of paragraphs will be written, it is important that *some* sort of end-of-paragraph marker be inserted at the time the material is written. It really doesn't matter too much what this end-of-paragraph marker is, so long as it is something that is not being used in the text itself. In one typesetting system that I am familiar with, the end-of-paragraph marker is the "at" sign, symbolized as @. However, any unique ASCII character not used in any other way could be inserted as an end-of-paragraph marker, and the front end could do a global search and replace with its own chosen symbol. If the typesetting system to be used is known at the beginning of the project, that typesetter's end-of-paragraph marker can be used.

Obviously, the end-of-line indication that is seen on the word processor screen has no meaning for the typesetter, as the spacing will be quite different. Thus, the author need not be concerned about where lines end on the screen, and short or long lines can be used as needed. It is frequently desirable to start a new sentence on a new line, even if the previous line is fairly short, as this makes certain editing operations easier.

In addition to the end-of-paragraph marker, any font changes within the paragraph and any special characters should be inserted at the time the material is typed. Again, the commands used need not be the ones that the typesetter will recognize, so long as they are unique and are not used in some other context as text itself. Whenever a word or phrase is to be set in some other font, such as bold or italic, it is necessary to insert a command into the text calling for the new font, and then another command at the end of the word or phrase calling for return to the normal font. These commands should be convenient to use and should be easily recognizable within the body of the paragraph. I use the slash, /, as the flag for an upcoming font change, and have followed it with a "b" for bold or an "i" for italic. Return to normal is then signalled by an "n" for normal font (some workers use "r" for "roman", which is the term used for normal text fonts). Spacing around font change commands is sometimes a little tricky, and the author should ensure that such things as periods and commas are placed on the desired side of the font change command.

Special characters placed within paragraphs present difficult problems for this simplified markup. Special characters frequently used in textbooks are degree signs, Greek letters, mathematical symbols, subscripts and superscripts, and commercial symbols. Each of these is called in a different way in the typesetter. There are two ways of handling special characters. One approach is to establish a command that will flag the typesetter operator to a special problem, and then insert the instructions on the hard copy of the manuscript. When the front end computer reaches this location, it will stop processing text and flag the operator, who can then examine the hard copy, see what the author had in mind, and insert the proper command. With this system, the only thing the author needs to know ahead of time is the flag character that will be used (a character should be selected that is not frequently used, such as the right or up arrow). Another approach is to define mnemonic codes for frequently used special characters and then embed these in the text, telling the typesetter exactly what codes have been used. Thus, a "degree" symbol could be coded [dg], a "times" sign [ts], etc. These mnemonic codes must be distinguished from real text by the use of special delimiters, such as curly brackets { } or "greater than" and "less than" < > symbols. It is essential that the delimiters used for mnemonic codes are not used as straight text.

Some other text elements that will be formatted separately by the typesetter and which should be marked in some way in the copy include section heads (both major and minor), lists of items (both numbered lists and lists in which each item is preceded by a special character such as a bullet or box), footnotes, quotations (usually called "extract"), and bibliographical entries. A whole series of markup procedures are required for typesetting tables. Although an interested author can insert markup for all of these elements (and many more), it is probably preferable to forego most of this markup and simply flag the material. There are so many variations in how these text elements are typeset that error is likely if markup commands are inserted. After all, the author is not being paid to be a typesetter.

Transferring the machine-readable material to the typesetter

Once the book is written and finally revised, with the minimal typesetting commands inserted, the material must be transmitted to the typesetter's front end. There are several ways in which this can be done. The most preferable, from the author's viewpoint, is to simply give a set of the floppy disks containing the text to the typesetter (or the production editor), and let the typesetter do the transfer and conversion. This requires that the disk format used is one which the typesetter can read. Most typesetters that handle material of this type are able to read disks in a variety of formats, so there should be no problem. However, if the microcomputer system used was a nonstandard one, or if the typesetter lacks conversion capabilities, then it may be necessary to transfer the material over a telephone line. We discussed communications capabilities earlier in this chapter. Most large typesetters are able to receive machine-readable text over the telephone through a modem. There are still some potential compatibility problems, but they can generally be overcome.

For transfer of text over the telephone, communications capability is needed on the word processor, which is generally available as an option on most systems. Communication requires two items, a hardware device called a *modem*, and a piece of software that will work with the modem to handle the text transfer. If this equipment is not available, it may be possible to obtain the use of a compatible computer which does have communication capability. You will have to tie up the computer for several days for this task, since the transfer of a book-length manuscript over the telephone is a fairly time-consuming process.

Summary

We have discussed briefly in this chapter the great advantages of doing a book using a word processor or microcomputer, and have strongly urged that this approach be used for a new project. Before serious writing has been done using a computer, it is important to select a suitable piece of equipment and the necessary word-processing software. The software and hardware selected should be sufficiently sophisticated that a large project can be done. A major requirement of the system is that it be able to handle large files and lengthy pieces of text. A system suitable only for letter writing and mailing-list preparation should be avoided.

The ideal arrangement is for the author to enter the text directly at the keyboard, probably from rough handwritten copy. Revision of the rough draft is made at the time of text entry. This eliminates the need for a secretary to retype the material, an expensive and time-consuming process. One of the major advantages of this direct-entry approach is that the tedious and costly process of proofreading is simplified. Once a word is spelled right, it remains spelled right in the computer. To catch the occasional misspelled word, spelling programs are available that will check each word in the text against an extensive dictionary of properly spelled words.

On a major textbook project, the publisher should be willing to pay for the use of the word processor, since the cost is really only a minor part of the cost of production. There are many advantages to the publisher if the author uses a word processor, most significantly the increased accuracy and reduced time required to put the manuscript in final form for production.

Another advantage of using a word processor or computer for the writing of a book is that it makes subsequent revisions much easier. Assuming that the book is successful in the market place, new editions will be appearing every four or five years. Some chapters will need to be extensively revised, but other chapters may well serve without major revision. If the text has been saved on disk, it is a simple task to generate the revised version.

We have seen that the text of the book, captured in machine-readable form by a word processor, can be transferred directly to a typesetter, and entry of the material again at the keyboard of the typesetter is not necessary. This has tremendous advantages for both publisher and author. From the publisher's viewpoint, the time required for typesetting the book is greatly reduced, and a major savings in cost is realized. From the author's viewpoint, proofreading of galley proofs is greatly simplified, since word-for-word reading is no longer necessary. It can be firmly stated that the computerized typesetter will not make small errors, such as generally are seen when the material is entered fresh at the keyboard. If errors do occur, they will be major ones, such as the production of complete gibberish, or massive distortion of the way the text is displayed. However, the typesetter is unforgiving of errors already present in the copy it receives. Any errors that have not been caught will of course be set into type.

We have seen that an ideal arrangement, when preparing material for the typesetter, is for the author to include some simplified markup commands directly in the text. Especially paragraph endings and font changes within the body of a paragraph should be incorporated by the author at the time the book is first written. Other markup, such as special characters, footnotes, section heads, etc., may be inserted, but alternatively can be flagged and left to the typesetter.

Before the author begins to do a book using a word processor, and especially before the text is prepared in a form which might be interpreted by a typesetter, a financial agreement should be reached with the publisher. It is clear that with the procedures described in this chapter, the author becomes the typesetter's keyboard operator. Typesetting is expensive, and is a cost always borne by the publisher. The author must be compensated for doing this expensive job for the publisher. It is not uncommon for the typesetting on complex material such as is found in many textbooks to cost $15-$20 dollars a page. For an 800 page book, this means a typesetting budget of $12,000 to $16,000. If the material is supplied to the typesetter in machine-readable form, the cost may be reduced by as much as 25%. Thus, a savings of $3,000 to $4,000 may be realized. Unprompted, the publisher may be quite willing to let the author go uncompensated and incorporate this savings into the profit of the company, or in a reduced price for the book. Certainly this is unfair to the author, who should be compensated for the effort and intelligence put into this complex job. The whole cost of the word processing equipment can be covered by the savings. The financial arrangements that the author can make with the publisher for providing the text of the book in machine-readable form should be incorporated into the initial contract that is signed. At any later time, the chance of getting significant financial consideration from the publisher may be minimal. We discuss the contract in detail in the next chapter.

9

The Contract between author and publisher

If the textbook is successful, the book contract between author and publisher is one of the most important documents the author will sign. A book contract has some similarities to a marriage contract. Just as a marriage contract ties husband and wife together for life, so can the book contract tie author and publisher together for life. However, although husband and wife are equal partners in a marriage (at least under present-day Western law), author and publisher are not equal partners in a book contract. The book contract almost always gives the publisher major advantages over the author. One reason, of course, is that a spouse is a human being, capable of being dealt with on an individual basis, whereas a publisher is a corporation, with huge financial resources and constantly changing personnel. The person signing the book contract for the publisher is not likely to be the person executing its terms, even if this person remains with the company throughout the life of the contract.

Stories abound of the contractual problems which authors have had with publishers. The paper by Baumol and Heim cited in the bibliography provides some examples from the academic world, and the magazine *Publishers Weekly* frequently mentions examples from trade book publishing. The Author's Guild also has published many examples in its newsletters, and presents a thorough discussion of how a good contract should be written. Publishers, of course, can also recount their own stories of pesky and cantakerous authors, but the point being made here is that the publisher has resources with which to handle such problems (as well as lots of experience), whereas the author generally does not. The following quotation speaks volumes:

> The asymmetrical relations between authors and publishers are symbolized by the publishing contract. Publishers still account to authors in their own way and frequently at their own pace. As in Islamic law, where men can divorce their wives at will but no such right is accorded women, a publisher has the right to refuse to publish a book even after a publishing contract has been signed. In contrast, an author cannot "divorce" his or her publisher. Every standard contract contains a clause specifying that the publisher will accept for publication a manuscript only when the publisher finds it satisfactory. No contract ever specifies that an author can terminate a publishing contract if he or she finds the publishing house's services to be unsatisfactory . . . The author can discover that he or she has spent several years working for no pay. *Coser, Kadushin, and Powell: Books, The Culture and Commerce of Publishing, p. 229.*

The goal of the author in negotiating a contract is to anticipate every problem and try to get it in writing in the contract. It should be strongly emphasized that oral assurances are useless. A friendly acquisitions editor who says (probably even in good faith): "Of course, we'll do *that.* There's no reason to clutter the contract with *that.*" may not be working for the publisher when the time to act is at hand. Get everything **EVERYTHING** in writing. At a number of places in this book, when discussing production matters, I have mentioned things that should be included in the contract. The point of the present chapter is to explain in detail how contracts are signed.

This chapter does not propose to give legal advice. In essence, the message of this chapter is "consult a lawyer before signing a contract". However, most lawyers are not very familiar with the publishing business, or with the implications of various clauses in publishing contracts. Most of the legal writings on book-publishing contracts deal with trade books. A textbook project is quite a different matter, so that the legal guidance available from a lawyer is often inadequate. Thus, the author can work better with the lawyer if the problems are understood sufficiently so that they can be discussed. Thus, the lawyer should be contacted *after* reading this chapter.

The standard contract

Most publishers have a standard printed contract. A model contract is given in the Appendix. First authors will often accept it as presented, and this is what the publisher hopes for. Here is the key point: **The standard contract is just a basis for negotiation. Do not sign it!**

The Author's Guild has published its own standard contract, which is available to members, and can presumably be consulted by lawyers. However, no publisher would sign a standard Author's Guild contract and, indeed, such a contract is frequently inappropriate for a text book. Examples of standard publishing contracts can also be found in legal textbooks. The definitive work on publishing contracts is Lindey (see bibliography), but that work deals only with trade book contracts. Although standard contracts are useful to read, in that they cover all of the essential elements of a publishing contract, they lack specifity. In this chapter, I go through in some detail the clauses of standard publishing contracts and explain how I think the author and the author's lawyer should treat them.

The publisher's view of the contract

The publisher has certain goals in signing a contract with an author. The publisher is interested in establishing a time when the manuscript will be completed and the length and general nature of the manuscript. Since the author is not an employee of the publisher, the publisher has no "power" over the author except in so far as it is spelled out in the contract. A major consideration of the publisher is that the manuscript be satisfactory. If the publisher signs with one author to do a book in a certain area, it is unlikely that this publisher will sign with another author to do a similar book. Thus, by signing with one author the publisher is eliminating the possibility of publishing certain other authors. The publisher is rightly concerned that the manuscript be satisfactory, and the standard contract always has a clause relating to this matter. Another matter related to the satisfactory manuscript is the problem of copyright infringement, libel, invasion of privacy, etc. The publisher will insist on assurances from the author that make the author responsible for copyright infringement, etc. Copyright matters will be discussed in the next chapter. It should be understood here that although the copyright in the book may be in the author's name, by signing a contract the author has transferred the copyright, or some of the rights arising from copyright, to the publisher. Thus, unless otherwise stated in the contract, when the author signs the contract, most of the rights to the intellectual property which represents his or her creative work are no longer in the author's control.

A major concern of the publisher is financial, related to royalties and to the division of moneys potentially available from the sale of subsidiary rights (on textbooks, primarily translations). The manner in which royalties are to be calculated, the frequency of distribution, and the procedures for accounting are spelled out in the contract.

The author's view of the standard contract

Because the standard contract is written by the publisher, most of the author's goals will not be included. The author is interested in avoiding an unreasonable rejection of the work. The satisfactory manuscript clause, which seems so reasonable when explained from the publisher's viewpoint, can be completely unreasonable if it is used to reject a manuscript because the financial situation of the publisher has changed since the contract was written, or if the editorial goals of the publisher have changed. The author is also rightly unwilling to accept *all* responsibility for copyright infringement and libel, if the author has taken great care to avoid such matters and a litigious reader has nevertheless taken offence at the author's use of material.

Financial considerations loom large in the author's mind. Although the publisher's employees get paid as they work, the author only gets paid after publication, and the amount paid may depend significantly on how hard the publisher's employees work to sell the book. The author is rightly interested in some guarantee that all the work has not been in vain. Although the royalty rate specified in the standard contract may seem munificent to the publisher, it may not seem that way to the author, especially if sales are not very good.

The important thing to bear in mind is that an author's ability to exact a departure from the publisher's usual form will depend on the strength of the bargaining position, i.e., on how eager the publisher is to put the author under contract.

Clauses in the contract

We discuss here briefly some of the key clauses in the contract, with explanations for their significance. At the top of any contract is a blank in which the author's name is inserted. In the case of multi-author works, the names of all of the authors will be included and it is understood that all of the constraints of the contract apply to each author individually. The name of the work is stated, and its approximate length. One of the major imprecisions of the book contract is that the actual work is specified only in terms of its name and length, with no indication of its literary content. Since virtually all contracts are signed before the book is written, one could rightly wonder how a court of law could ever interpret the contract in terms of the final book.

Grant of publication rights

A key clause that is near the beginning of the contract is that in which the author grants to the publisher sole and exclusive right to publish the work in book form. In the broad form, all publishing rights to the work are granted to the publisher. This means the right to publish and sell the work not only in the English language, but also in any other language. A significant right that may be specified here is the right of the publisher to issue a paperback edition. In addition, all subsidiary rights are also granted, such as television and movie rights. For textbooks, these subsidiary rights may be of little consequence, but if the book has potential in these areas, then the author should insist on retaining those rights. In trade book contracts, the original hardback publisher is frequently only granted the rights to publish the book in English and in hardcover, the translation, paperback, and other subsidiary rights being retained by the author.

For translations, the standard agreement is that the author and publisher split equally any proceeds arising from sale. If the author has some special reason to want to retain the translation rights (perhaps because of the international character of the text), then that should be explicitly stated in the contract. Generally, profits from translation rights are not large for textbooks, so that if an author has a legitimate reason for retaining these rights, the publisher may not resist.

Delivery of manuscript

A delivery date for the manuscript will be given. This date is established by means of discussion between the author and the acquisitions editor. The time lapse between signing the contract and the delivery date will probably depend upon how far along the project is. In the textbook field, where authors are not writing for a living but to supplement income, the completion date is rarely met. The manuscript may be delivered a year or more late, or never. A common observation in the textbook publishing business is that only 50% of the contracts signed ever result in published books!

If an advance on royalty is provided in the contract, then there will be a clause stating that the advance must be returned to the publisher if the manuscript is not delivered as specified, or if the manuscript is not satisfactory to the publisher. Depending on the size of the advance, the publisher may nor may not take legal action to bring about a return of an advance on an unfulfilled contract. Although the "satisfactory manuscript" clause is a major bone of contention in trade book publishing, where advances are often large, it is likely to cause little financial trouble in the textbook publishing field.

The author should realize that if the manuscript is already completed at the time of contract signing, more bargaining rights may be available to the author. The publisher may well be willing to forgo some of the traditional subsidiary rights and may be willing to pay a greater advance on royalties, if the manuscript is actually in hand.

Author's warranties

An important clause in the contract, from the publisher's point of view, is that clause which states that the author is the sole owner of the work and has the power to copyright it. This clause also states that the work does not infringe on any copyright, and contains no libelous or scandalous material. The clause goes on to state that the author will indemnify the publisher against any loss, expense, or damage due to breach of these warranties. The legal language in this clause is especially complete, because the publisher wants to ensure that it will not be damaged by any lawsuits resulting from the actions of an unscrupulous author.

Generally, this clause does not offer any problems to authors. If the author has not copied the work from others, there is no reason to fear the consequences. This clause does give the publisher a major escape route if the author uses material from others without permission and it points up the necessity of obtaining permission for the use of any figures or tables, and for giving full credit for anything taken from others.

The author might want to consider requesting that an additional sentence be added to this clause to circumscribe the extent of author's liability in case of suit. A clause sometimes added states that the author's liability shall not exceed the aggregate sums received by the author from the publisher. This protects the author against the (very rare) situation of an unusually litigious person who has found offense in something that the author has written. Publishers generally have liability insurance that protects them in case of a legal suit. The author probably cannot afford to have equivalent liability insurance, but many trade book publishers are now agreeing to include author's liability in their own insurance. A textbook publisher may well agree to this also, although the author may find the chances of a suit so remote as to not be worth any major discussion.

The Author's Guild provides very detailed language in its standard contract that offers additional protection to authors. For instance, the Guild recommends a sentence stating that the indemnity shall not apply to any material inserted into the work by the author at the publisher's request, a very reasonable sentence in a textbook contract, where significant editorial refinement may be done by the the publisher. Another sentence in the Author's Guild contract states that the indemnity shall not apply to any material which the publisher could have determined, from a reading of the work, clearly violated copyright or was libelous. This protects the author from poor editing on the part of the publisher, and places some of the responsibility for the integrity of the work on the publisher.

Agreement to publish

A standard clause specifies that the work shall be published at the publisher's expense. Language such as the following is used by one major publisher:

> If the Author fails to deliver a satisfactory manuscript on time, the Publisher will have the right to terminate this agreement and to recover from the Author any sums advanced in connection with the work. Until this agreement has been terminated and until such sums have been repaid, the Author may not have the work published elsewhere.

A desirable addition to this clause would be language specifying what would happen if the publisher does not act expeditiously to bring out the work. Language such as:

> If the publisher fails to publish the work within [one year] [eighteen months] after the final manuscript has been submitted, all publisher's rights shall revert to the author.

This sentence will protect the author against a publisher who becomes financially strapped or goes bankrupt, or operates so poorly that efficient production cannot take place. Manuscripts have been known to languish, either because of financial exigencies or gross inefficiencies of the publisher, to the significant detriment of the author. A one-year delay in publishing represents a major financial loss to an author on a successful textbook. An additional sentence that could be added to the contract, if a royalty advance has been provided, is that in the event the publisher defaults, any advances made by the publisher to the author shall not be returnable. The "satisfactory manuscript" clause gives the publisher an out, and it is reasonable for the author also to have an escape clause of this sort.

Publishing details

Although not common in trade book contracts, an important set of clauses in textbook contracts relate to the manner of publishing and the author's control over the final product. One standard textbook publishing contract has the following language: [The text in bold face is not in the printed contract provided by the publisher, but should be added by the author.]

> The Publisher will have the right to edit the work for the original printing and for any reprinting, provided that the meaning of the text is not materially altered. **Changes in the work must be approved in writing by the Author and Publisher.** The Publisher will have the right: (1) to publish the work in suitable style as to paper, printing, and binding; (2) to fix or alter the title and price; (3) to use all customary means to market the work. **The design, paper, printing, and binding of the work shall be subject to the Author's approval.** Author will deliver the manuscript in typewritten form. The manuscript will be submitted in duplicate and a third copy will be retained by the Author. It will be in proper form for use as copy by the printer and the content will be such as the Author and Publisher are willing to have appear in print. [If a word processor or microcomputer is used, and the author is providing machine-readable copy that will interface to a typesetter (see Chapter 8), then this should be specified at this point. The precise arrangements, including any payments to the author for providing machine-readable copy, should be specified.] The Author will read the proofs, correct them in duplicate, and promptly return one set to the Publisher. The Author will be responsible for the completeness and accuracy of such corrections and will bear all costs of alterations in the proofs (other than those resulting from printer's errors) exceeding [10%, 20%] of the cost of typesetting [This assumes that the author does not provide machine-readable copy]. These costs will be deducted from the royalty payments due the Author.

The 10% figure for author's alterations given in the paragraph above is often used (sometimes even 5% is used), but the publisher may be willing to change this to a higher figure. Since author's alterations may be a major cost (as much as several thousand dollars) on a first edition, the higher figure has some significant value. A useful addition here that will protect the author against faulty bookkeeping on the publisher's part is the clause:

> provided that an itemized statement of such charges is forwarded to the Author within [thirty] days after date of publication

Additional clauses can usually be added regarding publishing details, specifying things that the author deems important. For instance, the following clauses were added to one contract that I am familiar with:

> The Publisher agrees to: (1) assign the project to a special projects staff [names inserted] for production; (2) produce a book utilizing second color throughout and including up to 32 pages of full color; (3) employ a copy editor to be named by the Author, subject to the Publisher's approval; (4) employ an artist to be named by the Author, subject to the Publisher's approval.

Another set of sentences specify what items the author is to supply. In addition to the manuscript, the key items for textbooks include the index, a teacher's manual or answer key (for questions given at the ends of chapters), originals for halftone illustrations, and sketches for linedrawings. If any of these items are to be prepared at the publisher's expense, this should be so specified in the contract. For instance, the index is a major effort in a textbook (see Chapter 6), and the author may want to request that the cost of preparing the index be born by the publisher (this is especially advantageous if computer technology is to be used in preparing the index). A teacher's manual is often a chore to prepare, and rarely generates any significant financial gain to the author. Certainly, if the textbook itself has a large number of difficult questions, the author has the responsibility of providing an answer key. However, a teacher's manual is often considered to be much more than an answer book, since it may contain details on lecture procedures, sources of audio-visual material, potential exam questions, etc. The author may well be able to induce the publisher to pay for the cost of preparation of the teacher's manual.

Halftone illustrations are a major feature of most textbooks, and can entail a major expense. In science and technical books, where halftones are not just window dressing but convey content, the author must have complete control of the halftone selection and use. However, publishers include photo research as a cost of the book budget, and there is no reason why these costs should be born by the author, or deducted from royalties. My recommendation is that a clause such as the following be considered:

> Complete and final copy for all halftone illustrations shall be provided by the author. The cost of obtaining these halftone illustrations shall be born by the publisher, up to [specify amount].

The cost here could well amount to a few thousand dollars. If the publisher does not wish to specify a dollar figure at this place, then this cost can be included in a general expense figure specified in a later clause (see below).

A matter of principle here is that no items should be required by the publisher that involve out-of-pocket costs to the author that are not repaid by the publisher. Unprompted, many publishers are quite willing to let the author bear the cost of many publishing details. Since the publisher incorporates all its own costs for the production of the book into the final price, there is no reason why the author's costs should not also be included.

Subsidiary rights

Although frequently a major consideration in trade book contracts, where movie and television rights may provide more money than the book itself, subsidiary rights are of minor importance in textbooks. The following is the paragraph present in the printed contract of one of the major textbook publishers:

> The Publisher may permit others to publish, broadcast by radio, make recordings or mechanical renditions, publish book club and micro-film editions, make translations and other versions, show by motion pictures or by television, syndicate, quote, and otherwise utilize this work, and material based on this work. The net amount of any compensation received from such use shall be divided equally between the Publisher and the Author. The Publisher may authorize such use by others without compensation, if, in the Publisher's judgement, such use may benefit the sale of the work. If the Publisher itself uses the work for any of the foregoing purposes (other than publishing), the Author will be paid 5% of the cash received from such use. On copies of the work or sheets sold outside the United States, the Author will be paid a royalty of 10% of the cash received from such sales. On copies of the work sold through any of the Publisher's book club divisions or institutes, or by radio, television, mail order or coupon advertising direct to the consumer or through any of its subsidiaries, or to Elementary and Secondary schools, the Publisher shall pay to the Author a royalty of 5% of the cash received from such sales. If the Publisher sells any stock of the

work at a price below the manufacturing costs of the book plus royalties, no royalties shall be paid. All copies of the work sold and all compensation from sales of the work under this paragraph shall be excluded in computing the royalties payable under paragraph [number] above and shall be computed and shown separately in reports to the Author.

Although most authors will not be concerned with the restrictions in this paragraph, for certain textbook projects significant changes in language may be desirable. For instance, translation into foreign languages may be a major factor in textbooks in certain fields, and significant compensation may accrue. Frequently, the publisher is not sufficiently knowledgeable about the foreign market for a particular work to know how best to arrange translation, or to make any real effort to get the work translated. Translations are frequently done only if a foreign publisher contacts the U.S. publisher and expresses an interest. Thus, the book has had to first develop a reputation overseas sufficient to raise the interest of foreign publishers. If the book is one of the major textbooks of the year for a publisher, some effort may be made to market it for translation at the Frankfurt Book Fair, the main site at which foreign rights are arranged.

A common arrangement on translation rights is for the foreign publisher to pay the U.S. publisher a royalty, usually 7–8%, of its net proceeds. The lower royalty reflects the considerable cost in translation, and will cover the translator's fees. All costs of publication are born by the foreign publisher, although illustrative material may be provided at cost by the U.S. publisher (these would be taken directly from the negative film of the book, or from the camera-ready copy). Except for certain languages such as Spanish or Japanese, the market for translations of U.S. textbooks is not high, and no major income can be expected from this source. However, if the textbook has strong overseas interest because of its content or the reputation of the author, then the author may well want to retain all translation rights and arrange to market them. Depending on the work, many publishers may be willing to give up these translation rights.

The other sentences in the above paragraph relating to mail order and high school sales deal mainly with the lower income anticipated by the publisher. Mail order sales involve greater selling expense than traditional textbook sales, and the lower royalty is expected to compensate. Book club sales involve higher discounts than traditional textbook sales and this is supposed to be reflected in the lower royalty rate. However, since royalty is calculated based on net proceeds to the publisher (see below), it is not clear that these lower royalty rates are justified. The lower rate for overseas sales presumably reflects the increased cost to the publisher of selling the book in these far-flung markets, and the lower rate may be justified. If an author feels that the book in question has significant potential in any of the areas specified in this paragraph, then it would be very desirable to have specific advice from a literary agent knowledgeable in the textbook field.

Revisions and Competing Publications

An important series of clauses found in many textbook contracts relate to revision and to the publication by the author of competing books. Taken separately, these two clauses appear relatively harmless, but together they are potentially disastrous, virtually a time bomb. The following clauses are taken from the printed version of a major publisher's contract:

> The Author agrees that during the term of this agreement the Author will not agree to publish or furnish to any other publisher any work on the same subject that will conflict with the sale of this work.

> The Author agrees to revise the work if the Publisher considers it necessary in the best interests of the work. The provisions of this agreement shall apply to each revision of the work by the Author as though that revision were the work being published for the first time under this agreement. Should the Author be unable or unwilling to provide a revision within a reasonable time after the Publisher has requested it, or should the Author be deceased, the Publisher may have the revision prepared and charge the cost, including, without limitation, fees or royalties, against the Author's royalties, and may display in the revised work, and in advertising, the name of the person, or persons, who revise the work.

A third paragraph, already discussed, has the language:

> The Author grants this work to the Publisher with the exclusive right to publish and sell the work, under its own name and under other imprints or tradenames, during the full term of the copyright and all renewals thereof . . .

Taken together, these last three clauses have the effect of wedding an author to a publisher *for life,* with no possibility of divorce. There is no way that the author can terminate this agreement, because the author has granted the publisher the right to publish the work in all revisions, and the publisher has the right to *require* revisions. The net effect is to grant the publisher the right to extend the term of the agreement indefinitely. The publisher can do so by requiring the author to revise and by copyrighting the revision. The author has no practical way of terminating the agreement if the publisher decides to proceed in the foregoing manner. The author could refuse to do a revision, but then the effect of the contract is that the author loses control of the book. In addition, it is not clear whether the author would get *any* royalties on a revision done by others.

Because the author is wedded to the publisher until the publisher decides to release the author, the other clause above, the anticompetition clause, has major implications. The effect of the indefinite extension would be to bind the author to the publisher for life and to deny the author the right to publish any competing works in the field. In legal terms, this is considered a restrictive covenant and there is some question of its validity. However, the reasonableness of such a covenant could only be determined by the courts.

How is a "competing book" defined? Who is to decide if a book is competing? Although an anticompetition paragraph might seem reasonable if it extended over the life of an edition (say 4–5 years), it seems unreasonable when extended over the life of an author. What books *can* the author write for other publishers? Suppose the original contract specified a book with the title "Economics". This is so general that it would be impossible to know whether another book by the author in the general field of economics was or was not a competing book. The publisher may feel the book is competing, whereas the author may not. If large amounts of money are involved, the publisher could take the author to court. Since most contracts are written under the laws of the state of New York, the litigation would occur in New York courts. New York courts have found in favor of publishers when such cases have come before them in recent years. In addition, few authors would be willing to risk the considerable financial resources necessary to contest such matters in court.

We thus see that a series of paragraphs, seemingly reasonable when taken separately, are potentially disastrous when taken together. What can be done to avoid these problems? A lawyer knowledgeable about publishing law should be preferably employed to help the author negotiate. To control the open-endedness of the contract, a paragraph should be added making it possible for the author to terminate the agreement under certain conditions. The ideal language would make the contract applicable only to the present edition, with the requirement that future editions would require new contracts. Whether a publisher would agree to such a paragraph would depend upon how much it desired to obtain the book.

Concerning the anticompetition paragraph, its main intent is to protect the publisher from an unscrupulous author who finds a better deal with another publisher and wants to switch. If the publisher has already done major editorial work on a book, it is reasonable that it would not want to lose this investment. To protect the publisher, and still provide reasonable freedom for the author, a sentence could be added to the anticompetition paragraph defining exactly what a competing work is. One sentence that I have seen added is:

> The Author agrees that during the term of this agreement the Author will not agree to publish or furnish to any other publisher any work on [name of field] for the two-year college, four-year college and university level that will conflict with the sale of this work. A conflicting work is one that is of similar length and degree of complexity to the [name of existing book].

Although this still has some potential difficulties, it prevents a publisher from tying the hands of an author for other scholarly or commercial activity. Presumably, the publisher is not any more interested in taking the author to court than the author is in going into court, so that the author retains considerable freedom of action.

Termination or discontinuing manufacture

A paragraph that appears or should appear in all contracts relates to actions prescribed when the book is no longer selling and must be taken out of print. One paragraph that I have seen reads:

> When the Publisher decides that the public demand for this work no longer warrants its continued manufacture, the Publisher may discontinue manufacture and destroy any or all plates, books, and sheets without liability to the Author.

This sentence says nothing about author's rights to the work or about copyright assignment. A reasonable addition to this paragraph, taken from Lindey, would read:

> In case of discontinuance of publication, all the rights herein granted shall revert to the Author, together with any existing property originally furnished to the Publisher by the Author or at his expense.

A further set of sentences should be added here to specify what will happen to unsold copies (can the author obtain them at cost?), art and other materials supplied to the publisher by the author (should be returned to the author), and plates, sheets, or camera-ready copy (could be made available to the author at some specified fraction of the cost to the publisher).

A further paragraph is sometimes added here to specify what would happen if the publisher becomes bankrupt or makes an assignment of property for the benefit of creditors. Under these conditions, the agreement between author and publisher should automatically terminate.

Arbitration

Although rarely included in printed contracts of publishers, a useful paragraph recommended by publishing lawyers deals with how disputes should be handled. Lindey has provided the following language:

> Any claim, dispute or controversy arising out of or in connection with this agreement, or any breach thereof, shall be arbitrated by the parties before the American Arbitration Association, and under the rules then obtaining of that Association. The arbritration shall be held in the city of [specify], State of [specify], unless the parties otherwise agree. Judgement may be entered on the award in any court having jurisdiction thereof.

Royalties

A major set of clauses in the contract relate to royalties, and to how royalties are disposed. The question of royalties is reserved for another chapter.

Author's Agent

To acquire assistance in obtaining the best contract, the textbook author might consider the use of an Author's Agent. An agent is an individual who serves as an intermediary between author and publisher, providing guidance to the author and negotiating on the author's behalf with the publisher. Agents are widely used in the trade book field and there are some some agents in the textbook field, as well. For their services, agents receive a percentage of the author's royalty, usually 10%.

The agent should be involved in more than just the contract negotiations. An agent should work with the author on the development of the manuscript, advising on approaches and organization, perhaps even on content. Once the manuscript is complete, the agent should take the responsibility for finding a publisher, and carry out the necessary negotiations. Once the book is published, the agent will be responsible for collecting all royalties and distributing them to the author after deducting the agent's commission. The agent may also assume responsibility for handling subsidiary rights, for providing permission to those wishing to use the author's material, and for handling copyright matters.

The interests of the agent are in general the same as the interests of the author, so that the agent should be trusted to obtain the best possible contract for the author. If the agent has also worked with the author on the development of the manuscript, a better-selling book might result, with increased royalty for the author. It would be difficult to tell, in an individual case, whether the agent's commission is returned to the textbook author in higher royalties, but certainly some agents have made contributions with major financial consequences for authors of best-selling trade books.

The Literary Market Place lists over 300 agents, most of whom deal exclusively with trade books, but there are several agents who either handle textbooks as well as trade books, or specialize in textbooks. There are two trade organizations to which agents belong, the Independent Literary Agents Association and the Society of Authors' Representatives, both of which are listed in Literary Market Place. Either of these organizations may be willing to advise an author concerning the selection of an agent. The textbook author would be well advised to make extensive inquiries before signing an agreement with an agent to represent a particular work.

The agreement between agent and author can be either spelled out in a separate contract, or it can be included as part of the contract that the author signs with the publisher. Even if the author and agent do negotiate a separate contract, the agent's responsibilities and rewards will be included as a clause in the publishing contract.

It is strongly recommended that if an agent is used, this agent be knowledgeable about the college textbook field. College textbook publishing is quite different from trade book publishing, and most agents are oriented primarily to trade books. The author should find out what other textbook authors the agent is representing and then make inquiries from several of these authors before signing an agent agreement.

Conclusion

While I have avoided giving any legal advice, I have shown in this chapter that the contract signed between author and publisher is an important document that has serious implications. The contract as initially presented to the author will almost certainly be strongly slanted in the publisher's favor. How much the publisher will be willing to modify the standard contract will depend upon how much the publisher wants the book.

It should be understood that when the word "publisher" is used in this chapter, it has a somewhat undefined meaning. The person with whom an author will be negotiating on the contract is probably the acquisitions editor, who will rarely be able to take responsibility for signing the contract. Thus, if the author makes certain modifications in the contract, the acquisitions editor will have to "check back" with superiors to see whether these things are possible. The person actually authorized to sign a contract for a publisher will be an officer of the company, whom the author generally never meets. If an agent is not used, one procedure would be for the author to hire a lawyer to handle all negotiations. This would possibly give the author some psychological advantages in the negotiation, although not necessarily. However, the expense of hiring a lawyer for the negotiations may be too great to be justified. At any rate, the **final contract** should be examined by a lawyer *before* it is signed. Although this may not completely protect the author, the examination of the contract by a disinterested party is extremely desirable.

One final encouraging note: The Copyright Law of 1976 (see Chapter 10) clearly states that the publisher has no rights to the work *except* those that are specified in the written contract.

It should be emphasized that written contracts or agreements can, at best, provide only a broad framework within which the publisher and author work. Mutual trust is essential, and it is extremely unwise to allow any adversary relationship to develop. This does not mean that all reasonable concerns should not be spelled out in the written contract. But it does mean that neither author nor publisher should permit the written contract to substitute for friendly and open personal dealings. In the long run, the publisher and author have similar interests, and will both only prosper if they work together. The Publishers Association of the United Kingdom has developed a Code of Practice for dealings between publishers and their authors and this code is reproduced in the Appendix. Although such a Code has apparently not be adopted in the United States, most U.S. publishers would probably subscribe to many (if not all) of the clauses. An examination of the U.K. Code will give the textbook author some insight into the kinds of items to be concerned about in any dealings with a publisher.

10
Copyright matters

The Copyright Law deals with one important kind of intellectual property, that resulting from the creative activities of an author or artist. Copyright law gives the creator a number of rights which govern the publication and exploitation of the work. In the previous chapter, we discussed the book contract. One consequence of the signing of the contract is that some or all of these rights are transferred to the publisher and it becomes the publisher's responsibility to worry about the disposition of the rights invested under copyright.

Why, then, should the author be concerned about copyright? If the publisher controls the rights to the work, it can dispose of them in any manner it chooses and the author is interested in assuring that the use of his or her intellectual property is done in a professional manner, and that those using it are not debasing the work of the author. A successful textbook will elicit a large number of requests for use of drawings or figures, or chapters may be used in anthologies. Under copyright law, the disposition of the author's work for these purposes is vested in the organization to whom the copyright has been assigned, generally the publisher. Some control over the transfer of these rights to others should be provided to the author by the publisher . A knowledge of copyright law will help in understanding how these rights are maintained and transferred, and what restrictions the author can rightly insist on introducing.

The U.S. Copyright Law

The Copyright Law of 1976 became effective on January 1, 1978. It overturned many of the stipulations of the previous law, and has had enormous impact in certain areas. One of the major provisions of the new copyright law deals with "fair use" of copyrighted material, and affects most prominently the restrictions on photocopying material. Another major provision of the new law deals with the term of copyright, which is generally longer than that specified under the old law. Any work copyrighted under the old law has a different term than works copyrighted since 1978. Most of the other specifications of the new law, such as "fair use", will probably apply to works copyrighted before 1978 as well.

A point that needs to be emphasized is that copyright exists from the moment in which a creative work is "fixed in tangible form". The work is subject to the copyright law even though it has not been registered, and even if the words "Copyright by" are not placed on the work. Although registration of the work in the copyright office is not required, it has some major advantages in terms of the extent to which an infringer may be found liable.

"Fixed in tangible form" is interpreted broadly. In addition to the standard manner of writing the work on paper, other tangible forms include floppy disk, magnetic tape, keypunch cards, microfilm, or optical disk. The work must be "original". Originality is not defined in the law but is left up to the courts. The work need not be unique, creative, esthetic, or ingenious to be original. Indeed, an idea itself is not copyrightable, but only the "expression" of that idea. Further, facts are not copyrightable, but only the way in which these facts are expressed in tangible form. Thus, a table of data can be taken from a copyrighted work, rearranged in another form, and used without risk of infringement.

Many of the disputes that arise in copyright law relating to infringement deal with literary matters, and are generally of little concern in textbook writing. However, authors compiling textbooks from other people's material (such as collections of readings) must deal with the infringement problem.

Definition of authorship

Copyright vests initially in the author. No problem arises if only a single author is involved, but the case of joint works, common in textbooks, is more complicated. In the case of a joint work, the authors are co-owners of the copyright. Note that a joint work is different than a compilation of works of several authors. In a joint work, the work is prepared by two or more authors with the intention that their contributions be merged into inseparable parts of a whole work.

Joint authors are called "tenants in common", each of them owning an undivided share of the entire work. This does not necessarily mean that each author owns an equal share, since the authors can divide the work unevenly if they choose. The following is an important point for textbook authors:

> . . . because ownership is theoretically undivided, any [author] can transfer rights in the work without consulting the others. (If he does so, he will have to account to his fellow authors for their share of the profits.) However, a transfer of that sort will not prevent any of the other joint authors from making an identical sale to someone else, and for this reason, purchasers usually insist on getting all of the authors to sign the contract. (Strong, W.S. 1981. The Copyright Book: A Practical Guide. MIT Press, Cambridge, MA.)

It thus seems clear that when two or more authors write a textbook and all of their names appear on the title page, they are joint authors. On the other hand, if an author's name appears only at the beginning of the chapters that this author wrote, then the work is probably not a joint work. In this latter case, the copyright vests in the author of each chapter and the publisher or compiler is presumed to have acquired only the privilege of publishing the work as part of that particular collective work, any revision of that collective work, and any later collective work in the same series. However, in cases of this sort, the publisher will usually insist that the author of each chapter make an express transfer of *all* rights in the work to the publisher.

An important type of authorship that arises frequently in collective works, and may be important in textbook writing, is that called "works made for hire". If an author is employed by an institution for the express purpose of writing a work or works, then the work in question is considered a "work made for hire", and the owner of the work is not the creator but the employer who hired the author. An employer is considered to be the owner of *anything* written by an employee "within the scope of his [or her] employment". Thus, when a professor writes a laboratory manual for use in a course that this professor has been assigned to teach, copyright probably vests in the institution rather than in the professor, unless a prior agreement has been made. The institution may, if it so chooses, assign all rights to the work to the professor for disposal as the professor sees fit, but this requires a *written* assignment by the institution. The following quotation is relevant here:

> An interesting example of the importance of control is the case of university professors. From time to time universities have tried to assert copyright in faculty lectures. The courts have not agreed, and the crucial factor in their thinking seems to be that universities quite purposefully do not control the style or content of lectures. This rationale would not, of course, apply to internal memoranda written for administration purposes. (Strong, cited above)

A category of works which may be treated as work for hire are commissioned works. A commissioned work is one in which an individual or institution specifically requests that the work be prepared. Such works might be considered works for hire if they fall in any of the following categories:

- A contribution to a periodical or other collective work.
- A translation.
- A work that is supplementary to the main work, such as a foreword, bibliography, index, illustrative map, or chart.
- A compilation, such as an anthology.
- An instructional text, such as a study guide.
- A test, or answer material for a test.
- An atlas.

For such works to be treated as works for hire, the commissioner and the creator must agree so in writing. Publishers frequently have a separate printed contract dealing with works of this sort which specifically states that the work is considered a work for hire. Note that the term "work made for hire" is a legal term and should appear in a contract if the intent is to make such a transfer. Among other things, a work for hire has a different copyright term than a work not made for hire (see below).

Duration and notice of copyright

The duration of copyright depends upon the authorship of the work. The regular term is for the life of the author plus 50 years. In the case of joint authorship, it is the life of the last joint author to die plus fifty years. These terms are different than those in the old law, which specified a term of 28 years from the time of publication, renewable for another 28 years. In a work for hire, the copyright lasts for 75 years from the date of first publication, or 100 years from the date of creation, whichever is shorter.

As stated earlier, copyright privileges begin at the moment of creation and do not require registration. However, registration is essential for the protection of some of the rights under copyright. An unpublished work need not have a copyright notice, but whenever a work is published by authority of the copyright owner, a notice of copyright must be placed on all publicly displayed copies. Publication has a legal meaning: Publication is the act of offering copies to the public.

> Submission of a manuscript to a publisher is a classic example of "limited publication," as it is called, for which copyright notice is not required and would, indeed, be inappropriate. In the academic community there is also a custom of circulating so-called preprints (manuscript copies of forthcoming journal articles) among colleagues for comment and discussion. This too would be a limited publication in most cases. But if you are making what you intend as a limited publication to someone who might otherwise assume he was at liberty to do with his copy whatever he likes, you would be well advised to place on the copy a legend to this effect: "This copy is for private circulation only and may not be used in any other manner." (Strong, cited above)

Publishers occasionally neglect to include a copyright notice on a work. This is an editorial oversight, arising because of the complexities of book production and the large number of people involved in the process. Each person thought someone else was going to specify the copyright notice. The author is well advised to include in the contract (see Chapter 9) a requirement that the publisher copyright the work. Oversights may still occur, but if it can be shown by the contract that the author required the publisher to provide adequate notice, then it can be proved in a court that the faulty publication was not made with the author's consent. The law reads:

> The omission of the copyright notice . . . does not invalidate the copyright in a work if . . . the notice has been omitted in violation of an express requirement in writing that, as a condition of the copyright owner's authorization of the public distribution of copies . . . they bear the prescribed notice.

Form of copyright notice

The copyright law is very specific about the form of notice for copyright. It must include the following three elements:

1 The symbol ©, or the word "Copyright", or the abbreviation "Copr."; and
2 The year of first publication of the work; and
3 The name of the owner of copyright in the work.

The notice must be fixed on the work in a manner such that it gives reasonable indication of copyright. On books, this is usually the reverse of the title page, but other locations could be used.

The name in the copyright notice can be either the author or the publisher. The publisher of a textbook will generally be willing to copyright the book in the author's name. Note that the use of the author's name in the copyright notice has no special significance; all rights in the work have been conveyed to the publisher by the author upon signing the contract (see Chapter 9), and the publisher can dispose of these rights as it sees fit.

Registration of copyright

An important point is that copyright exists without regard to whether the work has been registered. Registration of copyright is important, however, because it is essential to enforce an infringement suit. Registration requires the completion of an application form and the deposit of two complete copies of the published work. If the Register of Copyrights determines that the material deposited is copyrightable, a certificate of registration under the seal of the Copyright Office will be issued. The effective date of registration, which will be specified on the certificate, is the day on which the application and deposit was received by the Register of Copyrights.

If registration has not been accomplished, and an infringement occurs, a registration application can be made at that time and an infringement suit entered. However, no award for statutory damages or attorney's fees can be made for any infringement made before the effective date of registration. Thus, it is essential that the work be registered as soon as possible after publication.

Most publishers have a special office to handle registration of copyright, and have standard procedures for ensuring that all of the requirements of the copyright law are met. After all, it is in their best interests to do so. In the process of protecting their own interests, they are also protecting the interests of the author.

Infringement and fair use

The owner of the copyright has the right to dispose of any or all rights. However, if someone uses part or all of a copyrighted work without permission of the owner, this is called infringement. The copyright law does specify certain categories of uses which can be carried out without permission of the copyright owner: such a use is called a "fair use". Textbook authors are concerned with infringement and fair use not only because unauthorized use of their works may occur, but also because they often need to use works of other people and will want to avoid infringement. In the education field, fair use is commonly permitted for certain uses, but there are limits beyond which a use is anything but "fair".

An author will, of course, never appropriate material from any source and use it without giving proper credit to the source, whether or not the source is copyrighted. If the source is copyrighted, then the author must obtain permission from the copyright owner before using the material. For books, the copyright owner is always printed on the reverse of the title page. In most textbooks, the copyright owner is the publisher, although the book may be copyrighted in the author's name (all rights to the book having been transferred to the publisher by the signed contract).

We have discussed the procedures for obtaining permission to use copyrighted material in Chapter 5. Most publishers have a standard permission form which can be used to obtain permission to use material from other sources, and a book will generally not be released for production until all permissions have been obtained. Note that in the contract the author has specifically agreed to hold the publisher free of all liability from infringement on other people's copyrights.

There is virtually no situation in which use in a textbook of another person's copyrighted work would be considered fair use. Fair use only relates to noncommercial uses in which the use would not be considered in competition to the original work. The law makes some specific statements regarding fair use. For instance, in determining whether the use made of a work in any particular case is a fair use the factors to be considered include:

1 the purpose and character of the use, including whether such use is of a commercial nature or is for nonprofit educational purposes;
2 the nature of the copyrighted work;

3 the amount and substantiality of the portion used in relation to the copyrighted work as a whole; and
4 the effect of the use upon the potential market for or value of the copyrighted work.

Obviously, with such language, any specific case could only be decided in the courts. The American Library Association and many major universities have developed guidelines for what would be considered permissible copying in an educational context. In the following, some of the uses considered permissible are listed (although interpretation of the law may change with time, as it is tested in the courts, and the following should not be considered as permanently applicable). For research purposes, individuals may make single copies of any of the following for scholarly research or for use in preparation of lectures for the class:

1 A chapter from a book.
2 An article from a periodical or newspaper.
3 A short story, short essay, or short poem, whether or not from a collective work.
4 A chart, diagram, graph, drawing, cartoon or picture from a book, periodical, or newspaper.

Most single copying for personal use for research, even if it involves a substantial portion of the work, would probably constitute fair use.

For classroom use, certain additional criteria are necessary. Copied material such as those types listed above would be considered permissible for distribution to students without the publisher's prior permission provided that:

1 The distribution of the same photocopied material does not occur every semester.
2 Only one copy is distributed for each student, which copy must become the student's property.
3 The material includes a copyright notice on the first page of the portion of material photocopied.
4 The students are not assessed any fee beyond the actual cost of the photocopying.

In addition to these general restrictions, certain length requirements have been specified. A prose work may be used in its entirety if it is less than 2500 words in length. If the work exceeds such length, the excerpt reproduced may not exceed 1000 words, or 10% of the work, whichever is less. In the case of poetry, 250 words is the maximum permitted.

If a work is no longer in copyright, then permission to use it is no longer needed, although personal integrity dictates that the source of the work must be cited.

The textbook author is sometimes concerned about other's use of his or her material not so much from the viewpoint of unfair use for commercial purposes (the publisher will take care of this), but for uses that the author considers improper or unseemly. A successful textbook will find many imitators, and some will be so slovenly that they will prefer to use material from other books rather than to develop their own material. The problem arises mainly in relation to figures, problem sets, and other uniquely creative materials that are difficult to "borrow" except by direct copy. Lazy authors may use half-tones that took the creator many hours to develop specifically for the textbook. Should permission be granted if, indeed, permission is requested?

A general principle that might be developed here is the following: If the material to be used is something that is a unique aspect of the textbook, which the author spent considerable time and creativity developing, then permission to use it in other books should not be granted. If the material is something that has arisen from the author's own scholarly work, and is being used in this author's textbook in the same way as it would be used if it had been taken from the scholarly work of another author, then permission should be granted. If the material is something that the author has taken originally from someone else, and has obtained permission to use it, then the author should not (indeed, cannot) grant permission to use it, but should refer the potential user to the original source. (I am constantly amazed at the number of people who request permission from me to use illustrations from my books that I have clearly credited to others.)

Even with these general policies, complications will arise in deciding whether to grant permission. Further, the publisher may well grant permission to use material without any consultation with the author. A useful procedure would be to write directly to the permissions department of the publisher and inform them that *all* permissions to use material *must* be cleared through the author first. Oversights may occur, but if the publisher has a well-run permissions department, such a request should be honored.

The manufacturing clause

A troublesome restriction in the present copyright law relates to manufacturing requirements and importation. In order to protect the U.S. printing industry, Congress specified that copyright by U.S. authors will be lost if the work in question is printed outside of the United States and more than 2000 copies are imported. This restriction does not apply to foreign authors. This section of the law has been challenged and may eventually be repealed, but as long as it stands, the author and publisher must ensure that the actual printing of the book is done within the United States, although preparatory work such as typesetting, color separations, drawing, and making of films and plates can be done elsewhere.

Summary

Most copyright matters are handled by the publisher and the author need not be concerned. The crucial matter is that the author transferred all rights in the work to the publisher when the contract was signed. Even if the copyright to the work is in the author's name, the author has no legal rights to the work except insofar as those rights are written into the contract.

One clause that should be included in the contract is the reversion clause, specifying that if the publisher allows the work to go out of print, that all rights revert to the author. Some textbooks may be so current and dated that transfer of rights to the author is of little consequence, but on other books (for instance, a collection of readings), renewed interest in the work may arise, in which case the author will be in a position to profit, provided the copyright has been transferred. When the copyright is reassigned to the author, the author will receive a legal document from the copyright office.

The textbook author is generally most concerned about using copyrighted material of others. Permission to use copyrighted material in a textbook must be obtained, and the procedure for this was discussed in Chapter 5.

11
Royalties

What percentage royalty should the author receive? The answer, for an author, is as much as possible. (Bailey, The Art and Science of Book Publishing)

Royalty is one of the more interesting topics dealt with in book publishing. For many authors, it is what makes the whole enterprise worth doing. Although textbook authors rarely make a living from their books, royalty can be a significant fraction of their total income. A general philosophy should be: The textbook should be worth doing of and for itself, because it will be a contribution to the discipline and to the field of education. But the author should not work for nothing. A clever author, writing and helping guide to print an outstanding textbook, should be suitably rewarded financially. Here is an exciting thought: With the right kind of book, published at the right time, the author has the potential of becoming independently wealthy. Many authors of major textbooks make more money from their royalties than they do from their regular salaries. On the other hand, many textbook authors work for virtually nothing. They may have published the wrong book, at the wrong time, perhaps with the wrong publisher, or they may have not been suitably reimbursed.

Another important principle that has been mentioned before but should be emphasized here: Royalty is compensation for the author's time and energy in writing and producing the book. It is not to cover author's *expenses* involved in the book project. Legitimate author's expenses should be covered separately by the publisher.

There are many confusing details in specifying and calculating royalties. Although publishers are unlikely to cheat authors, they are quite willing to let authors work for less than they should be working for. Although the legalities of the contract were discussed in Chapter 9, the topic of royalties was reserved for the present chapter. We will be discussing not only how royalties should be calculated, accounted for, and spent, but how the contract language related to royalties should be written.

How royalties are calculated

Although royalties are calculated on the basis of how many copies of a book have been sold during a given time period, there are at least two different ways of calculating royalties, which may give significantly different outcomes. Royalties can be calculated as a percentage of the list price of the book, or as a percentage of the net proceeds that the publisher receives. Since books are sold by publishers at discounts to bookstores, the latter procedure results in a lower amount of royalty, unless the percentage factor used is adjusted upward.

Royalty as percentage of list is the standard procedure in the *trade book* field. The actual percentage, however, is negotiable, and depends upon how much the publisher wants the book. The Author's Guild, which deals primarily with trade books, recommends the following as minimum rates for adult trade books:

1 10% on the first 5000 copies;
2 12.5% on the next 5000 copies;
3 15% on all copies in excess of 10,000

These percentages are minimums and an established author may receive 15% for the first copies with step-ups to 17.5% and 20%.

Since the publisher determines the list price of the book, there are potentialities for dissatisfaction with even the best royalty percentage. The Author's Guild recommends that a minimum list price be specified in the contract, but it is unlikely that any textbook publisher will agree in writing to a minimum list price for a book at the time the contract is signed. Publishers will argue that it is in their best interest to price the book as high as the market will bear, and that the interests of the author and publisher are identical. This is approximately true, but leaves a lot of room for author dissatisfaction.

There are many other complications in calculating royalty on trade books which are beyond the scope of the present discussion. Such items that come into consideration are mail order, coupon, or television promotions, book clubs, remainders, and other ways of selling books that frequently result in unusually large discounts or heavy promotion costs which the publisher may try to have offset by means of a lower royalty.

In contrast to trade books, royalty calculations on textbooks appear deceptively simple. The standard procedure, used by the major textbook publishers, is to calculate royalty based on the actual cash received by the publisher, that is, as a percentage of net income. From the publisher's viewpoint, this approach is simple and understandable. Varying discounts to different markets do not have to be reflected in the royalty paragraph of the contract, since the net proceeds will depend upon the discount rate. For textbooks, the discount to college bookstores is about 23%, so that one can readily convert a royalty based on list to a royalty based on net.

Let us assume that a $30 book is to bring a royalty to its author of 10% of list, or $3.00. The percentage would have to be 13% if the royalty were to be based on net proceeds and the publisher's discount were 23%.

However, although the regular discount to a bookstore on textbooks may be 23%, the average discount to all dealers may be higher, a figure of 33% being not uncommon. With a 33% discount, the royalty percentage should be 15%, if the author is to obtain the same royalty as would be obtained at 10% of list. In the printed version of the contract used by some of the major textbook publishers, a figure of 10% of net is often given. This figure can and should be crossed out and the higher figure of 15% inserted. Most publishers will probably agree to 15% of net, if they really want the book.

One basis for use of different royalty rates is the different marketing and promotional cost of books sold through different channels. Marketing costs are much higher for a book promoted in single copies primarily through direct mail or space advertising than for a book promoted to professors as a textbook. There is considerable leverage involved in textbook sales, because a single adoption will result in the sale of many books. For this reason, 15% of net is reasonable for a textbook, and a lower figure should not be permitted.

Most publishing contracts have some clauses which are troublesome to authors trying to decide what terms are suitable. For instance, consider the following language, in the contract of a publisher which generally pays a royalty of 10% of list:

> *Royalty Limitations:* In the event of special sales (defined as sales made at discounts which exceed wholesale discount schedules to customers whose normal business is other than wholesale or retail book distribution), the Author's royalty will be at the initial royalty rate stated above based on the proceeds realized from such sales.

This means that for special sales, instead of paying royalty based on 10% of list, the author is paid based on 10% of net. Is this unreasonable? Possibly not, if the copies in question would not otherwise be sold. Conceivably, the market is such that the book is not selling as well as anticipated. Both the publisher and the author may be left with nothing for all of their work. Under these conditions, it may be reasonable to unload the work at a lower price (higher discount). But why should the *percentage* to the author be lower? The publisher's reasoning is that it has invested a lot of money in the production of this book, and stands to lose a lot if the books are not sold. By selling the books in a special sale, the publisher may be able to recover costs, and because of the lower price, it is realizing much less than it would otherwise realize. Although this may seem reasonable to the publisher, the author may want to ask: Why me? Why should I pay the burden by a lower royalty rate, when the publisher and I are in this together?

Consider another clause of this sort:

> Should the publisher at any time have damaged or unsold or return copies of the work on hand which are not salable on the usual terms, it may dispose of such copies, and if sold at or below cost, no royalty shall be paid.

The words "below cost" here are not precisely defined, but presumably "cost" here is "production cost", the cost of editorial and manufacturing. The list price of a book is often about 5 times the production costs, which means that for a $30 book, these costs may be $6. If the books are sold at $6 each, obviously books at $30 will not be purchased and the author will lose any chance of royalty. At $6, the publisher is recovering a lot of its out-of-pocket costs, and it might be reasonable to ask why the author shouldn't also be paid something?

A similar clause in a major textbook publisher's contract reads as follows:

> On copies of the work sold through any of the publisher's book club divisions or institutes, or by radio, television, mail order or coupon advertising direct to the consumer or through any of its subsidiaries, or to Elementary and

Secondary schools, the publisher shall pay to the author a royalty of 5% of the cash received from such sales. If the publisher sells any stock of the work at a price below the manufacturing costs of the book plus royalties, no royalties shall be paid.

Whether this clause can cause any significant mischief will be depend upon the nature of the book, and whether it is amenable to marketing through any of these channels. Again, the lower royalty is intended to reflect the greater expense involved in marketing through book clubs or by direct mail, but it is not clear why the author should not be compensated at the same royalty rate. Many of these sales may well have been made anyway at the regular rate, and by marketing the book through untraditional channels the publisher is simply taking royalty from the author. The lower rate for elementary and secondary schools presumably also reflects the more expensive marketing required to reach these channels. Certainly, the royalty for a textbook whose content is such that it would find wide use in such markets could be heavily affected by this clause. If the author feels that the book may be used in the El-Hi market, then a detailed budget for the book should be requested, showing the unusual marketing costs and explaining why the author should be penalized. To a great extent, this clause is in the contract because standard El-Hi books involve extensive editorial development by the publisher, and the author involvement is frequently much lower than on college textbooks. Because the publisher has invested a large amount of money in editorial investment, and the author has made a correspondingly smaller contribution, the royalty rate is decreased. Reasonable enough, but if the author is writing a college-level textbook that has the potential for a large El-Hi market as well, the editorial development by the publisher was probably minimal, and the author's involvement has been the major factor in the book's content. The 5% royalty rate is not justified, and should be eliminated.

Another difficulty with calculating royalty rates based on list price is that some publishers use a technique called "net pricing", in which no traditional list price occurs. A net price, the price at which the book will be sold to the bookseller, is established and the bookseller charges whatever it needs to meet overhead, or whatever it thinks the traffic will bear. If net pricing is done, then a single-copy price which is something like a list price may still be established for mail order or direct sales. If a publisher does use net pricing, then clearly the only way that royalty can be calculated is based on the publisher's net proceeds. In such cases, the 15% rate should be the minimum.

Some publishers, interested in obtaining a particular highly desirable textbook, may be willing to modify the royalty rate so that it increases on a sliding scale, depending upon how many copies of the book have sold. For instance, the royalty rate may be 15% of net for the first 10,000 copies, going up to 18% for the next 10,000 copies, and leveling off at 20%. The number of copies in the formula will depend upon how expensive the book production is anticipated to be and how many copies are likely to be sold. The rationale here is that the lower rate is used until the publisher's production costs have been paid, at which time a higher payment to the author can be justified. Most publishers are unlikely to suggest a sliding rate unless they are in a competitive position with other publishers attempting to sign the same author for a major book.

One useful piece of advice when negotiating with a publisher over royalty: Ask to see sample information on royalty rates and total royalty payments for similar volumes already in the publisher's list. This will give the author an opportunity of determining how much money is really involved.

When considering the amount of money likely to accrue on a textbook project, it is important to consider that the author may actually earn more on a successful book through increased compensation from his or her institution, than through the royalty directly. This is because publication of a well-respected book, widely used in the field, will enhance the author's reputation, placing the author in a better competitive position in the academic marketplace. On the other hand, one certainly should not *count on* being rewarded locally, since a great deal depends upon the resources of the institution and upon how promotions and salary increases are handled.

Royalty on foreign and subsidiary sales

Although subsidiary sales (television, movies, etc.) may not be significant for a textbook, foreign translations may be a major factor. Some books are widely used throughout the world, and the potential sales are great. The standard contract language regarding foreign and subsidiary rights states that any compensation received from such use shall be divided equally between the publisher and the author. The net compensation to the publisher may only amount to a few thousand dollars, in which case the distribution is not worth haggling about, but on a major successful textbook that has become well established, foreign sales may be very high.

How are foreign translations handled? In most cases, a foreign translation occurs only if a professor in a foreign country finds the book very useful and wishes to use it for students. The professor would contact a publisher in his or her country and suggest a translation. If the foreign publisher agrees that the book would be marketable, then the foreign publisher contacts the U.S. publisher and buys translation rights. The foreign publisher agrees to pay the U.S. publisher a royalty (generally based on a percentage of list price) based on the number of copies sold. The royalty rate is lower, to reflect the fact that the foreign publisher must pay for a translation, as well as pay for all of the normal production costs. For the use of line drawings and half tones, the foreign publisher pays the U.S. publisher a fee, based usually on the costs to the U.S. publisher of duplicating this material.

The royalty rate that the foreign publisher will pay is negotiable, and may depend upon how great a demand there is for the book, but is frequently in the range of 7.5% of the foreign publisher's list price. Thus, if the foreign publisher sells 10,000 copies of a translation at a price equivalent to U.S. $40, then the royalty would be $30,000, of which the author would get $15,000. Although this is reasonable compensation to both the publisher and author (after all, neither had to do a thing to earn this money), most books do not have large foreign sales, so that the amount of money involved is usually not great.

The author may be able, through international contacts, to encourage or promote the translation of the book. Indeed, the author may be in a better position to do this than the publisher. Many publishers have extensive international operations, and may have sales and marketing personnel in many countries, but these people are more involved in selling the publisher's U.S. list than in obtaining translations. A major occasion for selling foreign rights is the Frankfurt Book Fair, held in Frankfurt, Germany every October. With publishers from over 80 countries represented, U.S. publishers have an excellent opportunity to sell rights. However, if a U.S. publisher has published one hundred books that year, the chance that much effort will be given to any particular book is low. The author's international reputation and foreign contacts will go farther toward ensuring translation than anything the publisher can do.

The major languages for translation are not necessarily the languages where the market is the greatest. English is the second language in so many countries that translations frequently are not needed, and in fact may be discouraged by professors interested in having their students learn English. The major languages for translation are Spanish and Japanese, with Italian, German, French, and Arabic farther down on the list. If there is likely to be major foreign market for the book, then considerable care should be taken in writing the book to avoid too much "local color".

Royalty on paperback editions

Although paperback editions may be popular with students and professors, from a royalty standpoint there is no advantage to a paperback. Paperbacks must be priced considerably lower than hardcover editions, and do not cost all that much less to produce. The only way that paperback editions pay their way is if their sales are significantly higher, and if the sale of a paperback does not decrease the sale of the hardcover edition.

Some publishers have a lower royalty rate for paperbacks. This is standard in the trade book field, where the paperback rate is carefully negotiated, and where the paperback publisher is usually a different publisher than the one doing the hardcover edition. Note that in the book business, two kinds of paperback editions are distinguished, the "trade paperback" and the "mass market paperback". The trade paperback is frequently similar to the hardcover edition, but with paper rather than board covers. In some cases, a book may be published originally as a trade paperback and a hardcover edition may never appear. Trade paperbacks are frequently sold at the same discount as hardcover editions, although because of the lower price, the income to the publisher and author is lower. This reduced income per book is presumably made up by the increased income due to the much larger number of copies sold. The Author's Guild recommends that an initial royalty of at least 6% of list be established for trade paperbacks, with a step-up to 8% after a certain number of copies are sold, although they state that established authors sometimes negotiate rates of 8%/10% or 10%/12% or even more.

The mass market paperback is another matter entirely. The mass market edition is an entirely different book than the hardcover/trade paperback edition, of a different size and with different paper and binding. The mass market publisher is a specialist existing just to publish mass market paperbacks. The mass market publisher acquires the rights to the book from the hardcover publisher and prints its own edition. Mass market paperbacks are distributed through different channels (magazine distributors are primarily involved), and are sold in drug stores, grocery stores, and other nonbook outlets as well as in traditional bookstores. The standard approach to the mass market edition is for the hardcover publisher to sell a license to a mass market publisher which includes specified royalty rates (for instance, 4% of the list price on the first 150,000 copies sold, 6% for all additional copies). The mass market publisher will also pay the publisher a lump sum on signing the contract, generally an advance on anticipated royalties. This license is a contract between the hardcover publisher and the mass market publisher. In the author's contract, language must be inserted that states how the author and hardcover publisher will divide the proceeds from the sale of the license. There are two parts of this disposition, the percentage of proceeds payable to the author, and the dollar figure at which this percentage changes. For instance, the Author's Guild recommends that on the first $10,000 of income, the author and publisher should split 50/50, on the next $10,000 the split should be 60/40 in favor of the author, and on all income above $20,000 the author should receive 70% and the publisher 30%. Some authors can do better than this, with 60% from the first copy and dollar. A successful author may even be able to retain all mass market rights and deal directly with the mass market publisher (generally through an author's agent).

The mass market is not generally a significant factor in the textbook business, but in many fields, certain books will do well both as textbooks and as mass market books. If the topic and approach are of potential interest to mass market publishers, then the author should ensure that proper compensation is included in any contract. The standard language in many textbook publisher's contracts dealing with this matter runs as follows:

> The publisher shall have the exclusive right to exercise or to license third parties to exercise any and all rights in the work in all forms and media throughout the world. The net amount of any compensation received from such use shall be divided equally between the publisher and the author.

This language is presumably satisfactory for most textbook contracts, where mass market rights are of minor importance. However, if mass market rights *are* of minor importance, then the publisher may well be willing to sign these rights over to the author, who would then be free to market them. The main problem from the publisher's viewpoint with this arrangement is that the publisher does not want to be placed in the position of competing with itself through another arrangement made by the author. Certainly there is room for negotiation here with some textbook projects.

Advances on royalties

Most contracts provide for an advance payment to the author, either in a lump sum or in installments, against future royalties. The advance is an important benefit to an author for several reasons:

- It shifts part of the risk of the book from the author to the publisher.
- The size of the advance that the publisher will pay provides a clue as to how important the publisher thinks the book is. If the publisher will agree to only a small advance, a few thousand dollars, then it does not think the book is going anywhere. An advance of as much as $50,000 could well be commanded for a major textbook.
- Receiving money in advance, especially in installments, helps the author from a tax viewpoint, since it spreads the income out over several tax years.
- It is essentially an interest free loan.

What happens if, for some reason, the book does not get written, or the book written is deemed unacceptable by the publisher? The contract will state that under these conditions the advance must be returned. If the advance is a few thousand dollars, and the book is not published because the publisher rejects it, the publisher often does not request the return of advance. However, lawsuits can and have been entered into to induce errant authors to return advances.

Consider the following documented case:

> A major U.S. textbook publisher signed on January 30, 1961 a contract with an author to write a history of England, to be delivered by October 3, 1963. Advances against royalties were paid as follows: $1000 upon signing, an additional $1000 on October 1, 1962, subject to certain conditions. An additional $5000 in advance was paid up to March 1, 1965, plus a payment of $750 for secretarial services (this latter was not an advance, but an outright payment). The manuscript was subsequently rejected by the publisher because it was deemed unsuitable. The author acknowledged in a letter to the history editor on Nov. 4, 1966 that the rejected manuscript was unsuitable as a textbook. (This was a mistake for the author to do, even if it was an accurate representation of the facts.) In this letter, the author suggested alternate arrangements to fulfill the contract, and the publisher agreed to extend the due date for submission of an acceptable manuscript to July 15, 1969 and to permit the author to retain the advances. Thereafter an additional advance of $1500 in November 1968 brought the total advances to

$8500. The due date was extended further to October 1971, then to March 1972, August 1972, and finally to the summer of 1973. In a letter June 12, 1973, the publisher wrote the author terminating the agreement and requesting repayment of the advances.

The author did send the publisher a check for $2000 dated March 18, 1974, but no further payments. On July 15, 1978, the author wrote the publisher that his book had been accepted for publication by another major publisher as a trade book. This caused the author to change his mind about returning any further advances, apparently because he felt that the publication of the book by another publisher indicated that the book did have merit. What subsequently ensued was a court case, in which arguments revolved around whether a textbook was the same as a trade book, and whether the author had or had not agreed that the contract was terminated and the advances had to be repaid. Ultimately, the New York court found in favor of the publisher, and the author had to return the rest of the advances. The final judgement was January 7, 1980, almost 20 years after the initial contract was signed!

In point of fact, most publishers are not so persistent, and if only a relatively small amount of money is involved, they will not go to court to get it back. This does not excuse an author from the responsibility of returning the advance, since to keep the money without proper performance is tantamount to theft. Extenuating circumstances may exist but do not really provide an excuse.

On a major book project, a publisher may be willing to offer not just an advance on royalty, but an outright grant of money. This grant is viewed as an inducement to the author to sign the contract. Grants are not common, and mainly arise when several publishers are competing for the contract for a major book. A grant of $10,000–$20,000 might be provided to sweeten the pot and make one publisher's contract look better than the others.

Should the author be concerned about a large advance or a sizable grant? In the long run, it is how well the book sells in the market place that will really determine how much money accrues to the author. The key thing is not how much money the publisher will contribute up front, but how well it will produce and sell the book. A publisher with a small sales staff and a weak production staff may be quite willing to provide more advance money, but then fall down during production or sales. It is more important that the contract provide for outstanding production efforts than that it provide for sizable advances. This is not to say that the author should not request an advance. From a tax viewpoint, the largest advance that the publisher will give is desirable, but the author should not be misdirected by discussions of the advance from the more important questions of book production.

Expenses versus advances

It is very important, when negotiating a contract, to make a clear distinction between money provided as expenses and money provided as an advance against royalty. Publishers are much more willing to provide a sizable advance against royalty than they are a sizable expense budget, because the advance is ultimately earned back in royalties, whereas the expense money can never be recovered. However, from a tax standpoint, expenses should never be considered as an advance on royalty. An advance on royalty is taxable, whereas payment for out-of-pocket costs, documented with receipts, is not taxable.

One problem may arise when the author asks for money to buy a piece of equipment, such as a word processor or computer. This is a capital investment, and will have many uses other than for the production of the book. The publisher is understandably reluctant to provide money for an outright purchase of equipment that is not intended soley for the book project. If the author were to buy the equipment from personal funds, the purchase might be tax-deductible, albeit in installments over 5 or more years as the equipment depreciates. Thus, the publisher may be willing to provide only part of the cost of such equipment. Exactly how such arrangements should be handled will have to be the subject of negotiations between the author and publisher with advice to the author from a tax consultant.

In Chapter 10, we discussed the contract itself in some detail, and indicated that all reimbursement for expenses incurred during preparation of the book should be specified in the contract. It is worth reiterating at this point that any money that the author wants *must* be written into the contract, since oral assurances by a publisher's representative may be meaningless.

How do the royalties roll in?

Major textbook publishers generally provide an accounting and payment of royalties twice a year, in spring and fall. The spring royalty payment represents sales made during the July to December period and the fall royalty payment represents sales made during January to June. Because of the nature of the college textbook business, the fall sales are the greatest, so that the spring royalty payment is markedly larger than that in the fall.

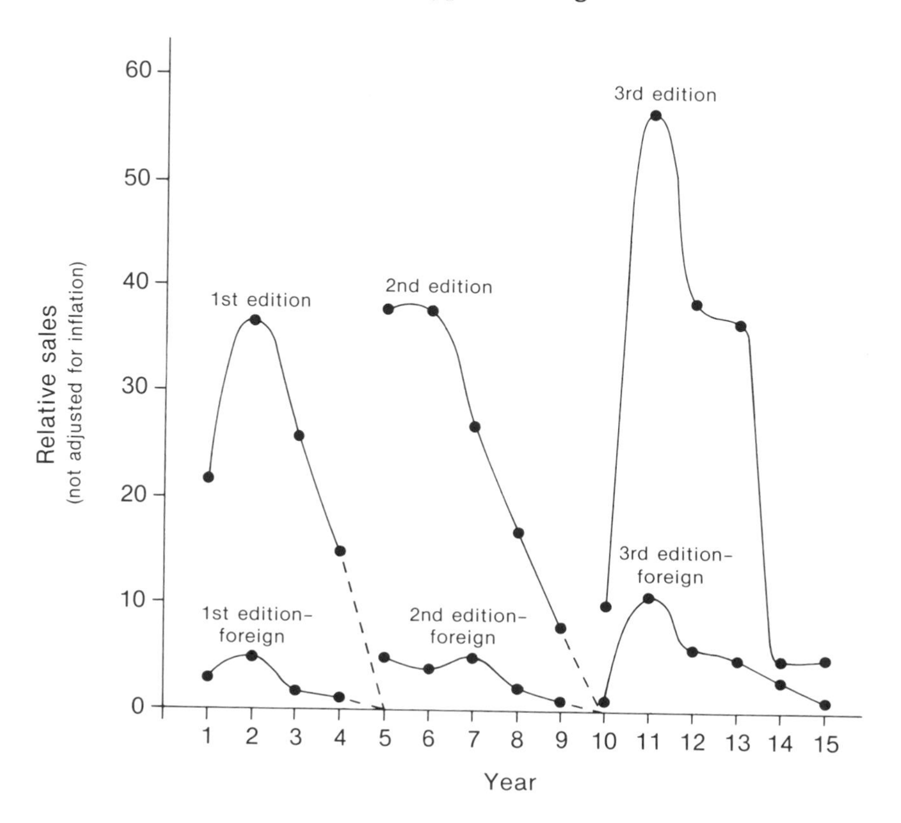

How can one determine that the royalty payment is correct? It is very difficult. It is possible to have a paragraph written into the contract specifying the possibility of an audit. Here is language that is sometimes written into contracts:

> The author shall have the right upon written request to have an accountant examine the publisher's books of account insofar as they relate to the work. The expenses of such examination shall be borne by the author, unless errors of accounting amounting to [specify percentage, such as 5%] or more of the total sums paid to the author shall be found to his disadvantage, in which case the expenses thereof shall be borne by the publisher.

What things are needed for a proper audit of income from a book? Only an accountant could provide a certain list, but the following things are among those that would be needed:

- Perpetual inventory records
- Price lists
- All tabulations of the book in sales reports by catalog number
- A listing of physical inventories with documentation as to how the inventories were verified
- Details of royalty rate computations
- General ledgers and books of original entry relating to sales, inventory, purchases, and manufacturing costs
- Company sales policies for the entire examination period
- Details of any movement of the book that has not been tabulated and does not appear on sales invoices

Obviously, an audit is only justified when a large amount of money is involved.

The cost of having an accountant in the publisher's location examine the books would be hard to determine, but it could conceivably be more than the author would be willing to pay. Publishers generally have their sales figures computerized, and the acquisitions editor that the author is dealing with will obtain a monthly printout of net sales. The author should be permitted to see these figures and thus be able to make an approximate calculation of the accuracy of the royalty statement. Note that raw sales figures are misleading, because they do not take into consideration returns, which on most college textbooks can be considerable (10% or more).

Royalty statements vary among publishers, and the amount of information given will probably depend upon how the royalty figure is calculated. If the royalty is calculated based on list price, and if there are several versions at different list prices (or both a trade edition at long discount and a text edition at short discount), then the royalty statement should give details of how many copies were sold in each version, and at what prices. If the royalty is based on the publisher's net proceeds, which is more common in the textbook field, then the royalty statement should state precisely how much money the publisher received in the various categories. For instance, income from foreign translations are distributed differently than income from the standard edition.

When the book goes out of print, hopefully at the time a new edition appears, there will be considerable returns of unsold copies to the publisher. The way most contracts read, the money reimbursed to the bookseller to pay for the returns will be deducted from the publisher's net proceeds. The author can well have a negative royalty statement for that period, for that edition. If a new edition has appeared, then the income from the new edition will be so great that the small amount of negative royalty on the previous edition will not be noticed. However, if a book is taken out of print completely, with no new edition, then the author may be in debt to the publisher. If the author has other books with this same publisher, the publisher may be able to deduct these negative royalties from the royalties of other books. For instance, the contract language of one of the major textbook publishers reads:

> The publisher may deduct from any funds due the author, under this or any other agreement between the author and the publisher, any sum that the author may owe the publisher.

This covers the publisher well, but the author may well wonder whether this is fair. If the royalties that the author has received are indeed unearned royalties (if the books are returned, they clearly have not been sold), then the author has really had use of some of the publisher's money. However, the approach specified here, of taking money from some of the author's other books with the same publisher, seems rather petty. Some publishers withhold part of the author's royalty payment to cover returns.

Controlling the tax problem

It is beyond the scope of this book to discuss in detail how an author of a successful textbook should control the income tax problem. If that desirable state is reached where the royalty income is as large or larger than the author's annual salary, the author might feel that tax problems are something that can be lived with. However, the royalty payments on a textbook last only for a short while. The largest competitor for new sales of the successful textbook are used books of the same edition. Once a book has been used at an institution for a year, there are large numbers of used copies floating around, which are sold back by students to a book store, and then resold to new students. Although recycling is meritorious, and the savings of money to impoverished students is salutary, used book sales make a significant dent in the author's royalty and the publisher's income. It is primarily for this reason, of course, that the publisher is interested in ensuring that a new edition is always coming along.

From a tax viewpoint, this boom or bust situation is undesirable. Because of our system of graduated income taxes, in a year when royalties are high, the taxes will be disproportionately higher. The procedure of income-averaging may help to reduce the tax burden, but only to a limited extent. Although authors of best-selling trade books have much more of a problem dealing with taxes than do textbook authors, tax-time still presents a problem for the textbook author. On a successful textbook, royalty income over the life of an edition might be as much as $200,000 (it is generally considerably less), much of which will come in the first and second years of the edition.

There are many ways of dealing with this problem, most of them complex. The Author's Guild from time to time provides advice to its members about how to deal with this problem, and the author will always find members of the investment community in his or her neighborhood willing to come forward and provide advice (for a fee!).

The key term that is used here is "tax shelter". The goal of any action is to shelter income from taxes. In general, the procedure is to salt away this money in some sort of tax-exempt form and keep it there until the author is ready to retire, and can then use the money. Since during retirement, the author's income is often lower, the tax bite will be proportionately lower.

Some authors handle this problem by incorporating. The Author's Guild has provided a good discussion of the advantages and disadvantages of incorporation as a means of sheltering income. If the author incorporates, then the author becomes an employee of the corporation, and a pension plan can be set up into which money is placed for investment. Since this money is not taxed, it is sheltered, and the income on this money is also sheltered. The publisher sends the royalty check to the corporation (which may have the same address as the author's home) rather than to the author directly. The pension plan money must go into a trust fund of some sort, managed by a trust officer (perhaps the trust department of a bank), and is unavailable to the author until retirement. Neither the pension plan money nor the income that it generates is taxed. On the negative side, the trust officer charges a fee for this service, which may be considerable. There is also a major cost of incorporation, plus the nuisance of keeping books and corporate records.

An additional advantage of incorporating is that all of the author's expenses related to this or other books can be paid out of the corporate income, rather than from after-tax income of the author. Thus, travel expenses for research, office supplies, postage, and other items not reimbursed by the publisher can be paid by the corporation. (If these expenses are directly related to the book, however, the publisher should be paying these as an expense. They should not be paid from royalties.)

A major disadvantage of incorporating is that if more than half of a corporation's income is from royalties, the corporation is considered for tax purposes to be a *personal holding company.* The profits of personal holding companies are taxed at a very high rate, and therefore the corporation must be certain to end the year without making a profit. This is done by paying out all income to the author, in the form of salary. Because the author is considered an employee of the corporation, not only must normal corporate taxes be paid (with correlative cost to the corporation for accountant fees), but unemployment compensation taxes must be also paid. Also, withholding forms must be filled out for both the federal and state tax authorities. Running a corporation is a nuisance, and is only worthwhile if the royalties on the book are so high that the net savings is significant.

Because of the fluid nature of the tax laws, nothing that is stated here can be applicable for very long. An author with the pleasant situation of excessive income taxes is advised to find a reliable tax lawyer or accountant knowledgeable in tax shelters.

Revisions after death

The book contract is an asset and becomes part of the author's estate. On a major textbook, royalty income may provide an author's family with a sizable annual payment. However, the textbook generally does not remain current for many years, so that new editions will have to be prepared. The contract almost always calls for periodic revisions of the work (see Chapter 9) and if the author is deceased then the publisher may obtain someone else to revise the book. The contract usually specifies that the cost of such revisions can be charged to the author's royalty. One difficulty with this arrangement is that there are no restraints placed on the publisher in obtaining the revision, so that most of the author's royalty may be dissipated in an expensive revision.

To avoid the loss of royalty to the author's heirs, the revision clause in the original contract should specify exactly what is to happen in the event of the author's death. One possibility would be to provide for author's royalties on revisions after death on a declining scale. On the first edition after the one the author prepared, the estate could receive 80 percent of the royalty, and over subsequent editions the percentage would go down.

Another possibility would be to require in the contract that the author's executor or author's spouse approve the choice of the revisor and possibly even approve the revision. This would provide some control on the publisher's activities in relation to that book. A third way would be to specify in the contract that in no event should the revision cost more than a certain percentage of the amount otherwise payable to the author, the percentage to be determined by how much work the author has done on the revision. Under all conditions, it is important that the author's estate share in the financial success of future editions, because it was the author's activities in the first place which made the book successful.

12

Marketing the textbook: the author's role

A successful textbook is made, not born. Even the best written and best produced textbook will not be successful financially if it is not properly marketed. Although marketing is the responsibility of the publisher, effective marketing requires author involvement. The publisher's staff needs the author's input because it is not knowledgeable about the fine points of the author's discipline. If the author provides this information, then the marketing staff will be able to target the promotional material for the book more precisely.

For someone not in the business world, marketing, sales, and advertising are often confused. **Advertising** deals with the production of advertisements, and includes copy writing, graphics production, printing, and distribution of advertisements. Two kinds of advertising are commonly used for textbooks: space advertising in journals, and brochures that are sent by direct mail to individuals. Much advertising is done by advertising agencies, who are hired by publishers either on a job basis or on a contract basis to prepare the advertisements. For space advertisements, advertising agencies are paid as a percentage of the billing by the medium, usually 15%. Thus, if a space advertisement in a professional journal costs $1000, the advertising agency receives $150, and the journal receives $850. The advertising agency also bills the publisher separately for the costs of preparing the advertisement, such as the graphic art and typesetting. Although some publishers may have their own internal advertising departments, they may still use advertising agencies for some of their work.

Sales deals with the actual sale of books to purchasers, and, in the case of college textbooks, obtaining adoptions. The sales department includes the field representatives and sales managers. Some publishers have very large field staffs, whereas others have small ones or none at all. A textbook can be marketed successfully without a field staff, using only space advertisements and direct mail, but most of the larger publishers are convinced that a field staff is the most effective way of marketing textbooks.

Since most publishers have a lot of textbooks coming out in any year, each acquisitions editor probably has one or two "big" books on which his or her reputation is riding. The acquisitions editor is paid partly on how well the books that he or she has signed sell. Thus, the acquisitions editor has a stake in getting the sales force out there pitching. The author rarely, if ever, deals with the sales staff, except on a casual basis. On the first edition of an especially important book, the author may be asked to attend a national sales meeting and talk to the field staff about the book. The point of such participation would be to fire up the sales staff and get them to go out and sell *this* book.

Marketing deals with the total activities related to getting a book to market. The marketing department is charged with developing a marketing plan for the book and making decisions on what types of approaches can be best used to sell that book. The author will deal initially with the marketing department, and will want to make a special effort to communicate clearly and concisely to the marketing representative.

The marketing department should be involved in the college textbook project from the beginning. Even before the contract was signed, the marketing department should have determined that a market sufficient to justify the publication of the book existed. Soon after the book has been launched into production, the marketing department should begin to develop a marketing plan for the book. The book budget will determine how much money can be spent on marketing. Here, the sales estimate enters the picture, because a book with a high estimated sales can justify a larger marketing budget than a book with a low estimated sales. On the average, the marketing budget for a college textbook runs about 14% of net proceeds, but many books receive more than 14% and some high-selling books probably less. If the first-year sales is predicted to be $500,000, a not unreasonable figure for a major college textbook, a $50,000 marketing budget could be expected. Although this may seem like a substantial marketing budget, it is not actually all that much money, since included in this figure are a large variety of costs. According to the Association of American Publishers, the two largest marketing costs are the cost of maintaining a sales staff and the cost of sending out complimentary copies. On a major college textbook, as many as 5,000 complimentary copies may be sent out, which could have manufacturing costs of $5–7 each.

A good marketing department will have an efficient program for the development of a marketing plan for the book. Often, the success or failure of the first edition of a textbook depends upon how effective the marketing program has been. The author can provide major help by communicating to the marketing department the relevant information in as simple and nontechnical a way possible.

The marketing questionnaire

Soon after the textbook is launched into production, the author should be sent a questionnaire by the marketing department. This questionnaire is a vital, even central aspect of the marketing program, since it provides the marketing department with the information needed to market the book successfully. An example of a questionaire is given in the Appendix. On a major textbook, a marketing person may actually conduct a personal interview, or an extended telephone interview, with the author. The author should remember that it is unlikely that anyone in the marketing department will be familiar with the discipline within which the textbook will be used.

The marketing staff should work initially with the acquisitions editor, who should have some familiarity with the market for the book. Correspondence from reviewers of the manuscript should be read, since these letters will provide clues to how the book is likely to be used. Feedback from sales representatives is often very important, since they are familiar with what books are being used in this discipline. The most important source of information on the book will be the author, who certainly knows the field and has some idea, perhaps inaccurate, about the market. Using all possible sources, the marketing staff will provide information that can be used by the advertising department to write journal and direct mail advertising, and will write the sales manual page that will be used by the field representatives.

The marketing staff, knowing nothing about the book or the discipline, is faced with the task of communicating information to other people (advertising and sales) who also know nothing about the book or the discipline.

What kinds of information will the marketing staff want to obtain from the author about the book? The following are some of the items that will be of interest:

- A short statement about what the book does and how the reader will benefit from it.
- A list of outstanding features of the book. This list should be specific, and should point out chapters that are particularly noteworthy.
- Teaching and learning aids provided within the book itself, such as questions at the ends of chapters, unusual illustrations, use of second color to highlight important points, marginal notes, glossary, etc.
- Supplemental material that will be provided to adopters of the book, such as an instructor's guide.
- Supplemental material which will be available for purchase by the students, such as a laboratory manual or a study guide.
- The table of contents, which can be shown to the professor by the field representative, or reproduced in advertising copy.
- Specific contrasts with competing textbooks, and why the present book is better than the competition. Names of specific competitors can be used here, so long as they do not make their way into printed advertisements (see below).
- Specific course needs which the present book is designed to fulfill.

An item that the marketing staff and sales representatives will find useful is a list of key terms that serve as highlights to professors of what is in the book, and a brief explanation of what these terms mean. If the book is the first book to provide detailed coverage of a major topic in the discipline, the field representatives will want to mention this, but they will not want to look like fools if they don't know what a word means.

A short biographical treatment of the author will also be useful to the field personnel, because this will help them build confidence among the professors in the qualifications of the author.

Once the marketing staff has assembled all the material and has conducted an interview with the author, the sales literature should be assembled. The author should be sent the final text of all material that is to be provided to the sales staff, as well as copies of all material to be used in advertising. The advertising copy will be written by the advertising department or by an outside advertising agency hired by the publisher, and should be sent to the author for approval before it is put in final form.

The marketing staff should make a special effort to obtain the names of specific college departments where the book might be used. College departments are often organized in unusual ways, or given nontraditional names (Life Sciences instead of Biology, for instance), and the sales staff may have difficulty knowing where a particular book belongs. Also, the name of the course for which a particular textbook is used may vary from place to place, and this kind of information should be requested. The level at which the textbook is aimed (freshman, sophomore, etc.) is also an important item of information.

Nontraditional markets for the book will also be considered. Would the book also be useful in high schools, preparatory schools, business schools, medical schools, technical institutes, adult education, training programs in government, industry or labor union? Will there be any professional market for the book, such as individual sales by direct mail or through book stores? Perhaps there are specific professions that might find the book of value, such as engineers, physicians, school teachers, etc. If the professor gives guest lectures frequently, it may be possible for the marketing department to have promotional material for the book for distribution at these lectures.

For advertising promotion, some detailed information should also be requested. Space advertising should be in journals or magazines that are seen by people likely to adopt the book, and the two or three best sources should be requested. A list of professional societies which might serve as sources of mailing lists for direct mail advertising should also be requested. Since some of the cheapest advertising is that derived from book reviews in professional journals and magazines, an extensive list of likely book review sources should be requested from the author.

Educational markets for the book

In addition to the information provided above, the author should give detailed information on the particular educational markets that this book will appeal to. College and university courses frequently have names that conceal their real intent. A biology course, for instance, may be called Biology, Life, Biological Sciences, Life Sciences, Man and Molecules, The Nature of Life, Man in the Biosphere, The Living World. Some of these courses will be for more advanced students, others for beginning students. Someone in the discipline will be able to figure out which course the book would be most suitable for, but the publisher's marketing staff may not be able to. Likewise, the titles of departments with similar missions vary widely across the country. A biology department may be called Biology, Life Sciences, Biological Sciences, or Natural Sciences, all of which have the same mission. In some institutions, there is no biology department at all, but separate departments of botany, zoology, physiology, microbiology, etc. In the larger institutions, departments that teach some aspects of biology may be called: Entomology, Physiology, Horticulture, Agronomy, Plant Science, Bacteriology, Physiological Chemistry, Biochemistry, Animal Science, Poultry Science, Forestry, Ecology, Wildlife Ecology, Environmental Science, etc. The publisher's field representatives are understandably confused, and may not know exactly where to market the book effectively. The more detail given, the better.

At the very least, the following information should be given about the educational market:

- A list of departments in which the book might be used.
- The titles of courses in which the book might be used.
- The level of course that the book is most suited for: Freshman, Sophomore, Junior, Senior, Graduate.
- The length of the course for which the book would be most suited: 1 Quarter, 1 Semester, 2 Quarter, 2 Semester.
- The time of year when such a course is usually offered: Spring, Fall, or both.
- Courses that are prerequisite for a course using this book. Listing prerequisites is very important. If the book requires mathematics, statistics, chemistry, or some other course as a prerequisite, knowledge of this permits the book to be marketed for the proper course.

In addition to the traditional college markets for the book, there may be nontraditional markets that should be contacted. Examples of these would be: high schools, preparatory schools, business schools, technical institutes, government training programs, industrial training programs, continuing or adult education programs.

In all cases, if there are particular individuals who would be most influential in the adoption of the book, the names of these individuals should be given.

Societies and Associations

Professional societies and associations provide some of the best information for the marketing of many textbooks. The author should provide detailed information on the appropriate societies or associations to contact. This should include the name of the organization, its location, address and telephone number, the name of a contact person (if known), and the mission of the organization. If the author has a membership list for the society, it would be desirable to send it to the publisher's marketing person. Some societies do not permit generation of mailing lists from their membership rosters, but many professional societies rent their membership lists to publishers for use in direct-mail advertising.

Mailing lists obtained from such organizations will be used to send either brochures describing the book, or sample copies. Most publishers agree that getting a sample copy in the professor's hands is the best way of selling a textbook. It is vital, however, that the *proper* person receive the book. Because of the fluid nature of college and university teaching, responsibilities for courses vary, and it may be difficult to get the book to the right person. The more information the author can provide, the better.

Direct-mail advertising is expensive but it is one of the best ways of reaching the appropriate individuals. Many small textbook publishers rely almost exclusively on direct mail to identify potential users. Those professors responding properly to a direct-mail solicitation will be sent copies of the book, and if the book is subsequently adopted for a class of 10 or more students the bill to the professor will be waved. However, the success of a direct-mail campaign depends upon the availability of an appropriate mailing list, and the author's information will be of great importance in guiding the publisher's marketing staff to the proper source.

Advertising in journals

An additional approach to selling textbooks is through advertising in journals (space advertising). Most publishers agree that space advertising does not sell many books, but it gets the name of the book out before the public. At the least, space advertising makes the author feel good, and informs the author's colleagues of this important activity that the author has been engaged in.

In order to use space advertising effectively, the publisher must be aware of the most suitable journals. The author should provide a list of these journals, in order of decreasing importance. A single space advertisement may cost $1000 or more, so that a wide use of space advertising is not possible. Because of the high costs, publishers may describe several noncompeting books in the same advertisement.

Review copies

In addition to the few journals in which space advertisements might be placed, there will be a larger list of journals to which review copies might be sent. Most professional journals have a book review section, and are seeking books to review. Sending out books for review is relatively inexpensive, since the books are already printed and no effort is required on the part of the publisher. The longer the list of suitable review periodicals that the author can provide, the better.

Exhibiting at annual meetings

One of the most effective ways of getting a new textbook before the likely users is by exhibit at annual meetings of professional organizations. The author should make a special effort to list the best meetings for exhibit of the book. Because exhibiting at annual meetings is expensive (in addition to the $800–1000 cost of renting the booth, there is the expense of shipping all the books and paying salary and travel expenses of the field representatives who will staff the booth), publishers select the meetings at which they will exhibit carefully. Exhibiting at small meetings is generally not cost-effective, but a large professional meeting, with 5000 or more in attendance, is an excellent place to market the book. The author should let the publisher know well in advance when the annual meeting is, its location, and the normal size.

Publicity on the book

Included in the category of publicity are activities such as announcement of the book in the communications media, author tours, autograph parties, etc. Textbooks are generally not promoted heavily in these ways, but for a particular book some of these activities may be suitable. If the author lectures widely on the subject with which the book deals, there may be a possibility of tie-in sales. Television talk shows are extremely effective ways of marketing trade books, but would be suitable only for the rare textbook. Most textbook authors are not inclined to get involved in this sort of activity.

Professional markets

In addition to the college and university market, many books may also be of value for professional workers, either established workers or those in in-service training programs. If there is a potential market for the book in these areas, the author should describe this in detail in the questionnaire. The author should explain why the professional person would want to buy the book and what major benefits the book has to offer. Are there valuable or unique reference materials in the book, such as tables, formulas, practical data, that could be applied directly to a professional reader's work? What are the titles of professionals who would most likely use the book. For example: chief engineer, sales manager, office manager, school principal, sales engineer, etc.

Marketing to these groups will almost certainly be done by direct mail, so that the availability of an appropriate mailing list will be crucial.

International markets

The larger publishers have extensive international activities, but virtually all textbook publishers attempt to market their books outside North America, either through their own efforts or through cooperative arrangments with foreign publishers. If a book has unusual potential for the International market, details of these markets should be given.

For English-language books, the most important international markets, in order of decreasing importance, are: United Kingdom, Europe, Australia-New Zealand, British Commonwealth, Middle East, Scandinavia, Latin America, Netherlands, Japan, Africa, and South East Asia. The Soviet Union and People's Republic of China are insignificant markets, even though they are very populous.

If a book is to be marketed overseas, it is essential that permissions that have been obtained for use of copyrighted material have no restrictions. The permission should be for rights throughout the world in the English language. If there are any restrictions on such use, then these restrictions must be cleared before the book can be marketed internationally.

The author should provide details of why the book would be of unusual use in these markets. Is there a significant global orientation, or does the book deal with material of unusual international importance? A list of countries in which the book would sell well should be given. If there are specific universities or colleges outside North America that should be contacted, these should be given. If the author has an unusual international reputation, this should also be stated. Because international marketing is expensive and difficult, any information that can be given that will permit more specific marketing of the book should be provided.

Annual and regional sales meetings

Sales information is communicated to the field staff by means of national and regional sales meetings. In the college textbook field, a national sales meeting is generally held in January of each year, to present to the sales force the new books and revisions which will be marketed during the main spring selling period for adoption for fall classes. At other times of the year, the larger publishers may also hold regional sales meetings to provide updates and more detailed information. It is at the sales meeting that the sales and marketing departments come together with the acquisitions editors to learn about the forthcoming books. At these meetings, it is usually the responsibility of the acquisitions editor to sell to the sales force the new books that are coming out under his or her jurisdiction that year. How well the acquisitions editor does the job may have a major influence on how well the book sells. This is because the sales representatives have a lot of new books to sell in any given year, and only a limited amount of time. How much time they give to a particular book may depend upon how fired up they were by the acquisitions editor's presentation. Hopefully, the author's acquisitions editor is effective at this part of the job.

The advertising copy

Advertising copy will be written for space advertising and for the brochure that will be sent out to prospective users. The author should have approval rights on all advertising copy. Some contracts actually include language to this effect. Most publishers of textbooks will clear all copy with the author anyway, but such contract language may be useful if the author has any doubt about how the publisher might advertise the book.

The advertising copy writers generally have little expertise in the field, and must rely on information provided by the author. The more useful the information that is included in the questionnaire, the better focussed and more effective will be the advertising copy. If a good slogan for the book comes to mind, do not hesitate to suggest it, as it may be better than anything the copy writer comes up with.

It is essential that the marketing and advertising people working on the book know that the author has insisted on seeing *all* copy before it is used. There are always time constraints at the advertising stage, and in the rush to get to the printer the author's approval may be overlooked. Some really embarassing copy is written by copy writers, but it will be the author who will look bad when the piece reaches readers.

Mentioning competing books

Although it is commonly done in the advertising world, the mention of the competition should be absolutely forbidden in the textbook field. Consider the following copy of a space advertisement which actually appeared (names have been changed to protect the innocent!):

> Jones/Smith/Brown is a better balanced introductory text than White and Connor/Brook put together. Because, unlike the competition, Jones/Smith/Brown covers all and *only* the basic aspects of microbiology.

Such an advertisement reveals the impoverished imagination of the copy writer. Even worse, it reflects on the authors, who should not have permitted such a travesty. (The book was unsuccessful, and is no longer on the market.)

Good copy writing is an art, but unfortunately most copy writers are trained in the mass market advertising fields, and have little feel for the college textbook field.

The author should look at a number of direct mail and space advertisements for college textbooks. The advertising department should be quite willing to put together a packet of material which includes advertisements for similar books, or for books at the same level in other fields. Check the quality of the pieces, paying special attention to the copy. Not only will this give a good idea of the quality of the publisher's advertising, but will give some ideas about copy that the author could write. If the advertising manager likes what the author writes, it may well be included in the advertisements.

Working with the marketing department

On the first edition of a textbook, the author will be contacted by a representative from the marketing department. If the book is considered one of the "major sales targets", then the marketing manager will spend a lot of time discussing the book with the author. The marketing manager may actually visit the author personally and spend several hours going over the key points in the book.

A good marketing manager will take pains to learn as much about the book as possible before the author interview. The marketer should read all the prepublication reviews, look through the manuscript or galleys, examine illustrative material, read the author's marketing questionnaire, and any other material provided by the author or by the acquisitions editor.

At the actual meeting with the author, the marketer will be obtaining specific sales information that will be of use by the field staff in their efforts to do an effective selling job. Any sales approaches or sales leads that might come out of the interview will be communicated to the sales department. The initial drafting of advertising or catalog copy will be discussed. Comparisons with competing books will also be made, and this will probably be one of the main points of emphasis.

When meeting with the marketer, the author should have available all significant material on the book, such as manuscript, galley proofs, page proofs, or copies of earlier editions of the book, or of competing books. The author should actually spend considerable time examining competing books, so that the strengths of the new book can be emphasized. The point here is not to downgrade competing books, but to outline in detail the strengths of the new book. The author should be encouraged to think about how the book will fill needs in the course for which it is intended. Some of this was of course considered when the book was written, but that was a long time ago, and many things may have changed.

Another important matter for discussion with the marketer will be the supplementary aids that will be available for use with the book, such as instructor's manuals, laboratory manuals, study guides, sets of overhead transparencies, 2 X 2 color slides, etc.

The author should remember that the marketer will have little or no understanding of the technicalities of the field. The terms will mean little, although there will possibly be some recognition of key words. The author should take special pains to explain things simply and clearly.

The author and marketer are working together to coordinate the marketing program for the book. The author should resist the temptation to cut the interview short, or to avoid the interview altogether.

It is important to do more than just list the key features of the book. The author should explain to the marketer why each feature is a key feature, and how the professor and student will benefit from its use. For instance, it is not sufficient to state that the book has an innovative format. It is necessary to say something like:

- This text's programmed format will give the students more laboratory time and will ensure faster comprehension of the basic theory.
- The boxed examples give the student real-life situations to consider without altering the flow of the main text.
- The special problem sets will give the students much more experience with the use of "X factor" and its application for contemporary society.
- The four-color illustrations explain for the first time the basic principles behind the theory of "Y".

Once the marketer has prepared the written material, it should be returned to the author for checking. The author should go over the material carefully, taking special pains to be certain that the material is accurate and factual. Remember that this material will serve as the basis for writing advertising copy and for the material which will appear in the sales manual. Errors here may be spread widely throughout the sales campaign.

The problem of effective communication to the market by way of the marketing department is a difficult and complex one. Whole books have been written on how to market books effectively. The more carefully the author works with the marketing department, the more likely the book is to be sold successfully. However, a key point is that although effective marketing can significantly increase the sale of a good book it is unlikely to permanently affect the sale of a bad book. The primary key to successful textbook publishing is to write a good book in the first place.

Sales

Most college textbook publishers employ sales representatives whose job is to contact professors and obtain adoptions. The effectiveness of a sales representative is often debated. Publishers vary considerably in the number of sales representatives employed. Some of the smaller textbook publishers have no sales representatives at all, and handle all sales contacts through direct mail advertising. A few of the larger publishers have over 100 sales representatives in the field.

Obviously, a large sales force is a major financial investment for a publisher, and the cost of the sales force must be incorporated into the price of books. Publishers with large sales forces must publish hundreds of new titles a year, and maintain large numbers of backlist titles, in order to be able to spread the costs of selling.

The sales representatives must be able to sell books across a large number of disciplines and generally must rely on the information provided in the sales catalog when they are attempting to obtain adoptions of new books. Some of the larger publishers have big enough sales forces so that some specialization is possible, and sales representatives may concentrate in general academic areas, such as science and technology, or humanities and social sciences. With the smaller publishers, the sales representative must be able to sell across all of the academic disciplines. Each sales representative will have a specific geographic territory, which hopefully will not be too large.

Sales representatives are paid in part on adoptions they have obtained, and the number of copies per adoption. Because of this, sales representatives will frequently spend most of their time at the larger universities, where heavy enrollments result in large sales numbers. The sales representative's job in obtaining an adoption is to make certain that the professor has a copy of the book, and to try to communicate why *this* book is preferable to the book currently being used. The sales representative should have the responsibility for maintaining an up-to-date mailing list of teaching faculty at the schools in the territory. It is essential that the new textbook be in the hands of those able to influence the adoption. This is a difficult matter, because university departments are extremely unpredictable in assignment of teaching responsibilities. Even the chairperson of the department may change so frequently that communication problems arise. One of the important functions of the sales representatives is to keep on top of teaching responsibilities for the courses with large enrollments, so that complimentary copies can reach the proper individuals.

The main way in which college textbooks reach the hands of students is through the college bookstore. There are a relatively limited number of college bookstores, and a single campus generally has only one or two. In some institutions, the bookstore is college owned and operated, whereas at other institutions the bookstore is a private organization. Both types of bookstore operate in essentially the same way.

Well in advance of the beginning of the semester, the college bookstore will contact each professor on the faculty and ask for a list of required and optional textbooks for courses that will be taught by that professor. Adoptions for the fall courses are generally requested in the spring, and adoptions for spring courses in the late fall. By the end of the final examination period in the spring, bookstores will have most of the information they need regarding fall textbook adoptions. Since colleges and universities begin new academic years in the fall, the most important time for solidifying adoptions is in the period from March to May. Rarely will a publisher bring out a major textbook later in the year than April or May. It is preferable to have the book out by late November or December, with a copyright date of the following year, since this will permit some adoptions for the spring term and also permit the professors to have plenty of time to examine the book for adoption the following fall.

When the college bookstore contacts the professor about adoptions, information will be sought not only on the course number and title, but also on the estimated number of students. Since professors are frequently optimistic about anticipated enrollment, and because there is frequently more than one bookstore serving a single campus, the bookstore will probably order fewer books than the professor requests. Books will go on the shelves for fall classes during the summer (August is the busiest month at a college bookstore). Because cost of books is a major inventory investment for the bookstore, any unsold books will be returned to the publisher as soon as feasible. The standard practice is for the bookstore to pay for all the books ordered, and then to be credited for those returned to the publisher within a reasonable period of time.

From the viewpoint of discount schedules, books are generally divided into two categories, trade books and textbooks, as discussed briefly in Chapter 11. Trade books are those sold to the general public, primarily as individual copies through bookstores. Trade book discounts will vary among publishers, but frequently run around 40–45%. Textbooks are sold at a lower discount, which generally is set at 25%.

The rationale for the lower discount on textbooks is that the bookstore is selling large numbers of copies, and has a fairly good idea even before ordering the book how many copies will be sold.

The net amount that the publisher receives from the sale of textbooks to a college bookstore is the list price less the discount. The bookstore will probably be expected to pay the shipping charges. Unsold books returned to the publisher in good condition are then credited to the bookstore's account, so that the net proceeds which the publisher receives from the sale of an individual book is calculated from the net price less returns. It is from this latter figure that the author's royalties are usually calculated.

The sales representative must work closely with the college bookstore to ensure that the needed books are available at the proper time. Because of the heavy weight of books, and the large numbers used, they must be shipped by truck rather than through the mails, and such shipping is slow and frequently unpredictable. If the books are not in the bookstore, students and professors will be angry, and damage to the reputation of the publisher and author may ensue.

Once a book has been used at a given college campus, a number of used copies will be available. Most students have no need for a book once the course is over. College bookstores generally have a used book department, where books are purchased back from students. Such books are then reshelved for sale to new students during the next term. The biggest competition for sales of a given college textbook comes from used copies of the same textbook. After three or four years, the number of used copies available is generally so great that sales of new copies are markedly depressed. The existence of the used book market is one of the main motivating factors for new editions. Once a new edition of a book is out, the used book market for the previous edition drops to nothing.

The price that a student receives for a used book from the local bookstore will depend to a considerable extent on whether this title is to be used again the following term. By the end of the spring term, college bookstores generally have most of the information about textbook adoptions for the fall term. If the book is to be used again, then the bookstore will buy the book back from the student at about 50% of list price (if the book is a hardcover book in good condition). The best time for the student to sell the book is during the period May 1 to August 31, when the bookstore has most of its adoption information and has not filled its stock completely. After August 31, the bookstore will pay the student less money for the book, because of uncertainty about whether the book will be used in the subsequent spring term. The value of a used book is determined by need. By December, many professors will have provided to the bookstore information about spring adoptions, and students can then sell their books at a higher price until just before the spring term begins. However, if the professor has selected a new book, or the book has been revised, then the used book is obsolete and is worth much less on repurchase. Under these conditions, the college bookstore will still repurchase the book, but at a much lower rate, and will then sell the book to a wholesale jobber. The jobber will then transport the book to another location where the book is still being used as a textbook, or will remainder the book. Because of the jobber's services, colleges beginning to use a textbook several years after it has been published may still have access to used books, thus continuing to cut into potential sales of new books. The only way that the publisher, and author, have of countering this competition from the used book market is by periodic revision.

International sales

One of the minor aspects of college textbook publishing is the international market. Although books at the professional level sell well overseas, college textbooks frequently sell poorly. Whether a textbook will have a significant international market may depend partly on the topic, and partly on the treatment of the topic. Textbooks at the freshman/sophomore level have less overseas market than textbooks at the more advanced level.

There are two aspects of the international market: direct sales of already published books, and rights for foreign translations. Most of the larger publishers have international divisions whose goal is to market books around the world. Minimal sales forces, or commissioned sales representatives, may be employed in many parts of the world. (A commissioned sales representative is not an employee of the publisher, but an independent sales person who sells the books of a variety of publishers, receiving a sales commission, generally 10%, on each book sold.) The primary direct sales market for college textbooks is western Europe and Japan. Since the book is written in English, it will find its primary use in English-speaking countries, or in those countries where English is an important second language (The Netherlands, Scandinavia, Germany, Switzerland). Western European countries frequently have college courses that resemble those in North America sufficiently that textbooks can be used without unnecessary difficulty on the part of the students.

Foreign translations are a minor part of the income from most college textbooks, but can be significant for certain books. The standard procedure is for the author and publisher to divide equally any income received from sale of foreign rights. Foreign rights may be sold outright for a flat fee, or may be sold for a percentage of the sales of the foreign translation. The latter arrangement is probably preferable for books with any reasonable anticipated sales. Because the foreign publisher has the expense of translation, the royalty to the publisher is lower than a standard royalty, generally 5-7.5% of the foreign publisher's net proceeds. Foreign translations may be a feather in an author's cap, but will almost never amount to much income. The languages into which college textbooks are most frequently translated are Spanish, Japanese, German, Italian, and French.

Reviews, good and bad

Finally, the book is out and the reviews begin to appear. The author should send up a strong cheer if the book is reviewed at all, since more books are not reviewed than are. Many journals have a policy of not reviewing textbooks, and even those that do review textbooks often provide only minimal coverage of a field.

Hopefully, when the reviews finally appear, they will be glowing, favorable, or at least neutral. Negative reviews are extremely disheartening and make one wonder whether all the sweat and tears were really worth it. What determines the quality of the reviews? Often, merely the "luck of the draw" in how a journal assigns book reviewers. The assignment of book reviewers is frequently a casual process, with little thought given to the selection of the appropriate reviewer for any particular book.

There are several things to consider about bad reviews: 1) Their effect on the author's ego; 2) Their effect on the way the author is treated by colleagues and fellow faculty members; 3) Their effect on sales. I will only consider the third point here.

The sales of a truly good book will probably not be hurt much by a single bad review, even if that review is published in a key journal. (An exception to this might be a bad review by an especially influential person.) However, if several bad reviews appear, this will probably cut sales significantly, and if *all* reviews are bad, the book is probably doomed. It should be noted, however, that many people do not read book reviews, and a bad book, even one unfavorably reviewed, may find considerable sales if it is targeted at a market in which there is little competition. Such a book does, however, have an uphill battle, a battle which it may never win. If all reviews are unfavorable, the book probably really is bad.

One way to reduce the likelihood of bad reviews is to have the book reviewed thoroughly and critically in manuscript. However, occasionally even a book which received good reviews in manuscript can garner bad reviews when it is finally published. Sometimes this is because of errors in production (figures with wrong captions, inappropriate or erroneous figures) that the author missed or was never allowed to see. Other times, it may be because the manuscript reviewers were uncritical, lazy, or inappropriate. Sometimes the author is just incapable of writing a good book (although this too should have been pointed out early in the manuscript-development process.

Bad reviews are especially depressing to an author who has done a conscientious job and knows the effort that has gone into the book. Sometimes the reviewer is unfair, ignorant, sarcastic, or inaccurate. Rarely in life do people receive such direct criticism as authors receive from book reviewers. An author who has published a book has laid a reputation on the line. The person writing the review has nothing to lose.

The publisher will, of course, be interested in seeing the reviews, good or bad. If the reviews are good, the publisher may use them in future advertising. If the reviews are bad, the publisher will welcome from the author an explanation. However, for the publisher the bottom line is not good reviews, but good sales. If the sales are outstanding, bad reviews will be quickly forgotten.

The fate of a successful textbook

We have now reached the end of the story. The book has been written, edited, produced, and sold. The author has worked well with the publisher, from acquisitions through editorial production to marketing and sales. The royalties have been generous and the book has been acclaimed a success. The author is now on the road to a successful part-time career and can anticipate continued success from future editions. The textbook will serve as a handsome supplemental annuity for retirement, and will enhance the author's reputation with colleagues. Students will be stimulated to enter the field and become professionals in their own right. All of the hard work and diligence has paid off. Congratulations!

13

The textbook author and the employing institution

The textbook author is generally a full-time employee of an institution. Publishing a textbook is generally not one of the normal duties of the employee. The author's activities thus fall in the general category of "moonlighting". However, doing a textbook is not moonlighting in the same way that working in an automobile repair shop would be. Because of the close, often intimate, relationship between work on the textbook and the normal "work" of an academic employee, problems of conflict of interest arise. Although each author's local situation will be different, there are some general concerns that all textbook authors should be aware of. These concerns are discussed briefly in this chapter.

Improper use of institutional time for work on the textbook

The general problem here is one of conflict of time. Faculty are expected to teach, to conduct scholarly research, and to publish the results of their research. Some aspects of textbook writing could be considered "scholarly", but most of it is not. If the author uses institutional time for writing the textbook, it could well be considered a conflict with the author's normal duties. On the other hand, writing a textbook certainly improves the author's teaching ability, and if the textbook deals with subject matter that is part of the author's teaching responsibilities, then the author acquires greatly improved teaching skills.

The problem here is to draw the line between work that will merely make money for the author, and work that will enhance the author's institutional responsibilities. Each situation will be different. The point is for the author to be sensitive to the nuances, and to discuss the problem with colleagues, department chairs, and (if necessary) deans.

The approach least subject to criticism or misunderstanding is one in which all significant work on the textbook is done in the author's home, during hours that would normally be considered the author's private time (early morning, late evening). Institutional time might be used for incidental matters such as obtaining references from the library, making telephone calls, and handling occasional correspondence. The justification for such institutional use is based on the fact that most professors use "personal" time for institutional activities, such as grading papers, reading and preparing for lectures, having meetings with students. Thus, overall, there will be a balance between private activities on institutional time and institutional activities on private time.

Although the activities of faculty members are usually not monitored, the author should behave as if they were, and carefully avoid using institutional time for major work on the book.

Conflict of interest

Although conflict of time, discussed above, can be handled by judicious attention to where and when certain activities are carried out, conflict of interest is a more vexing matter. Here we are concerned with the potential lack of objectivity of a faculty member.

The mission of a faculty member in an institution of higher learning is to discover and communicate knowledge. Credibility is maintained by complete objectivity, and by the knowledge that the faculty member is neither hiding things nor communicating things that aren't true.

With textbook publishing, the main conflict of interest arises from the improper use of the book by the author. If the author teaches the course for which the book is suited, then it would seem natural that the book would be used in the course. However, there will be the perception by some that the author may be requiring students to use the book not because it is the best book, but because the author obtains royalties from the sales. The ethical dilemma here is not handled by turning over royalties received from local sales to the institution.

In a large department, where many people are involved in teaching a course, the conflict of interest problem is less severe. If all who teach the course use the author's textbook, then this provides some objective credibility to the use of this textbook by its author. Also, if textbook selection is handled by a committee, then the author whose book is selected is less subject to criticism. Certainly a textbook which has found wide use across the country is less likely to be suspect than one which has very limited use, with the majority of the sales being at the author's own institution.

One way of avoiding a conflict of interest is to not write textbooks in areas of one's teaching responsibilities, but this seems a ridiculous solution. The author is best qualified to write textbooks dealing with his or her main teaching responsibilities.

One policy that I have heard stated at several of the major campuses is the "10% rule". The professor is not permitted to require students to buy a book that the professor has written if those sales would be more than 10% of the total sales of that book for the year. The principle here is apparently that if the book is widely used, the professor is not using it *just* to increase royalty payments. However, this 10% rule would certainly be difficult to enforce.

Publisher's field representatives generally assume that an adoption is assured at an institution if the author of the book is a member of that institution. Indeed, some commercial publishers exist just to publish books that are authored and used only at a single institution. If the enrollment is over 500 per year, then an inexpensively produced version can be economically viable if the adoption at this institution is assured. The 10% rule would serve to eliminate such obviously improper situations.

How to handle royalty payments

Although it happens only occasionally, royalties on a successful textbook which serves a large market can be handsome. It is even possible that the royalties on the book may be more than the salary that the author receives at the institution. Even a moderately successful textbook may generate an attractive royalty payment. Is this bad?

If the author has carefully avoided the conflict of time and conflict of interest problems just discussed, then there should be no problem about receiving and keeping royalty payments. The author is being paid for the expertise, the extremely hard work (outside of normal working hours, to boot), and the intelligence involved in the textbook project. The money accruing to the author is no different than money accruing from successful investments. In no way should any of that money be returned to the institution, and high royalty payments should certainly not be used to justify a lower salary for that professor. I have heard of cases where the royalties on those copies of a textbook sold at the professor's own school are returned to the professor's department, or to some endowment fund at the institution. I must confess that I do not understand the logic of this procedure. Without royalty payments, very few textbooks would be written, and students would be the worse for it. Why should the royalties generated locally be considered any different than those generated at other institutions?

A piece of advice that I could give to a successful author is to never, under any conditions, discuss with colleagues anywhere the size of royalty payments or the number of copies of a book sold. Financial disclosure is not a condition for employment in most commercial enterprises, and it certainly should not be a condition for employment at an educational institution.

Frequently, uncritical individuals confuse money received with potential conflict of interest. A conflict of interest does not necessarily arise because a large amount of money is involved; conversely, even with minimal or no financial involvement a conflict of interest could arise. Thus, financial disclosure is unlikely to help control conflict of interest situations. In the present context, all a financial disclosure would do would be to embarrass the professor and possibly permit the local institution to justify paying the professor less because of the high royalty payments that professor is receiving. The professor should not be paid an amount determined by need, but an amount determined by that professor's worth to the employing institution.

An important point to make, if the problem of royalty should come up, is that royalty payments are not a continuing source of income. As we have seen in Chapters 11 and 12, royalty is high in the first year or two of a textbook, but then drops dramatically. Unless new editions are brought out every four or five years, the royalty that an author receives will only serve as a significant supplement to a regular salary for a short period of time.

A successful textbook author will be less likely to be entangled in a difficult situation if he or she scrupulously avoids any discussion of royalty. Colleagues or friends will certainly try to draw their own conclusions about the size of royalty payments. Let them.

Copyright matters

We have discussed the copyright law in Chapter 10 and have noted that copyright vests in the creator from the moment that the work is placed in tangible form—that is, written on paper. The author can convey rights in the copyright to another, such as a publisher, and this is done by means of the contract, discussed in Chapter 9. The institution that employs the author on a full-time basis generally has no rights in the copyright, unless a specific contract between the professor and the institution regarding copyright matters exists.

Institutions may publish general policies for copyright matters, and at some institutions professors may be expected to sign a statement that they have read and will follow all institutional regulations. Faculty rules and regulations will vary in detail from one institution to another, but generally are fairly similar across the country. The professor, as author, should be certain that the local regulations do not specify retention of copyright by the institution.

Certain kinds of writing tasks fall under the general heading of "work for hire". The copyright law has specific regulations about work-for-hire arrangements. In such situations, the copyright does not vest in the author, but in the employing institution. Generally, educational institutions only specify work for hire for writing tasks that are clearly related to the duties of the employee, such as the preparation of tests, laboratory manuals, course outlines, etc. If a work has been prepared under a work-for-hire arrangement, the institution may still be willing to permit the author to publish this work commercially, but a specific agreement to that effect should be obtained. A textbook would not be considered a work-for-hire product unless a specific agreement had been signed *prior* to the time work had begun.

Generally, institutions are not in a position to publish and market books of a commercial nature. Although many institutions operate their own publishing house (the University Press), the mission of the Press is more to publish scholarly works that are unsuitable for commercial publishers than to exploit commercial works. In addition, most Presses try to avoid giving preference to local authors. Since most work-for-hire publications would be commercial rather than scholarly, and since the author would be local, it is unlikely that an institution will insist on retaining copyright to something prepared in this manner, and the author will probably be able to obtain a release. The work could then be submitted to a publisher for commercial publication in addition to the local uses for which it was originally written. It is preferable that the author avoid getting into a work-for-hire conflict in the first place, by carefully separating textbook writing from regular institutional duties. In case of doubt, the author should discuss the matter with a lawyer, or with the counsel for the employing institution.

The rewards of textbook writing

You are now aware that textbook writing is an enormously complicated and time-consuming activity. Textbook writing is certainly not for every one. Is it worth it?

A successful textbook will not only be profitable financially, but will also enhance the reputation of the author in his or her institution and throughout the country. A widely used textbook gives the author instant "name recognition", which (unfortunately) seems to be one way in which the reputation of an individual is assessed. Since the reputation of an institution is a reflection of the reputation of the professors in that institution, a successful textbook will generally enhance the reputation of the department or institution in which that professor is established. Such an enhanced departmental reputation provides a satisfying counterweight to any possible imbalance that might be construed as conflict of interest.

However, the greatest reward from writing a successful textbook is knowing that students' lives have been enriched. I have been greatly moved by students completely unknown to me who, many years later and now well established workers in my field, have told me that my textbook is what got them first interested in this field. Certainly, such statements, even if rarely given, are worth more than all the royalties that will accrue.

Most people do not tell other people how well they are doing. The author of a successful textbook may not receive compliments from colleagues and students, even if they value the book highly. But the greatest compliment of all is the adoption list that the publisher mails to the author from time to time. It is gratifying to learn that a book is widely used throughout the country, perhaps throughout the world.

Textbook writing is guaranteed to improve one's teaching prowess. The act of organizing a body of knowledge is of great benefit in course development and lecture preparation. Textbook writing will also probably improve one's scholarly research. Most research is done in extremely narrow fields, and applicable material from related fields is often overlooked. Preparing a textbook forces one to expand horizons, to look critically at other fields. Ideas arising from textbook writing will creep into one's own research in many direct or indirect ways. This has the potential for pioneering scholarly efforts. Textbook writing should not be viewed as competitive with scholary activity, but as a complement.

Should you do it? Obviously, textbook writing is not for everyone. It should never be done for money, and it should rarely be done for strictly local reasons. The point of doing a textbook is to provide for students and professors in your field a framework of reference for their learning and teaching. If you have the abilities, the time, and the energy, then go for it!

Bibliography

Association of American Publishers. One Park Avenue, New York, NY 10016. This trade association publishes occasional booklets on college textbook publishing, for distribution to potential authors. See for instance: *An Author's Guide to Academic Publishing* and *Books and Bucks, The Business of College Textbook Publishing,* both brief booklets. Also available is the AAP Greenbook of College Textbook Publishers (part of which is reproduced, with permission, in the Appendix to the present book).

Authors Guild, Inc. 234 West 44th Street, New York, NY 10036. This is an organization of authors. Although primarily oriented toward trade book authors, some textbook authors are members. The Guild publishes a Model Contract and a quarterly newsletter which provides much practical information on matters of interest to authors. Membership is only available to individuals who have published books.

Bailey, Herbert S.,Jr. 1970. *The Art and Science of Book Publishing.* University of Texas Press, Austin. 216 pp. This book contains an excellent discussion of book budgeting and planning. Although oriented primarily to the publisher, it provides many insights of value to the author. Unfortunately, published before most of the new technologies of book publishing were widespread.

Balkin, Richard. 1981. *A Writer's Guide to Book Publishing. 2nd edition.* Hawthorn/Dutton, New York. 239 pp. This book by a New York author's agent is oriented primarily toward trade book publishing, but provides many useful insights.

Baumol, William J. and Peggy Heim. 1967. *On Contracting with Publishers: or What Every Author Should Know.* American Association of University Professors Bulletin Spring 1967, pp. 30–46. Although financially dated, this article is required reading by any professor considering a book contract. Gives many practical examples of how the author can rewrite sections of the contract to ensure adequate protection.

The Chicago Manual of Style. 13th edition. 1982. The University of Chicago Press, Chicago. 738 pp. The "bible" of most editors, a must for any serious author. Especially good on copyediting and indexing problems.

Coser, Lewis A., Charles Kadushin, and Walter W. Powell. 1982. *Books. The Culture and Commerce of Publishing.* Basic Books, New York. 411 pp. This book provides the results of an extensive research study of the publishing industry sponsored by the National Institute of Mental Health. Although heavily criticized by many publishers, this book provides marvelous insights into the inner workings of the publishing enterprise. Of special value for textbook authors is the considerable coverage of how college textbook publishers work.

Council of Biology Editors. 1983. *CBE Style Manual, 5th edition.* Council of Biology Editors, Bethesda, MD. 324 pp. A widely used style manual for textbooks published in the biological sciences.

Dessauer, John P. 1981. *Book Publishing. What it is, what it does. 2nd edition.* R.R. Bowker, New York. 230 pp. A brief overview of the publishing business, with emphasis on trade books.

Goldberg, David. 1981. *Infringement of Copyrights.* Practicing Law Institute, Course Handook Series Number 134. Practicing Law Institute, 810 Seventh Ave., New York. 288 pp. One of several Practicing Law Insitute publications dealing with legal aspects of publishing. Although the reader may not want to delve too deeply into legal matters, many lawyers outside New York publishing circles may not be aware of these publications.

Henn, Harry G. 1979. *Copyright Primer.* Practicing Law Institute (for address, see Goldberg). 785 pp. This book provides the complete text of the U.S. Copyright Law, and an extensive discussion of each provision.

Kaufman, Henry R. (editor) 1981. *Book Publishing 1981.* Practicing Law Insitute (for address, see Goldberg). 880 pp. This book provides complete copies of court cases involving book publishing, examples of model contracts, and other documents of use to lawyers advising authors.

Labuz, Ronald. 1984. *How to Typeset from a Word Processor: An Interfacing Guide.* R.R. Bowker, New York. A practical guide to hardware and software for those interested in typesetting from microcomputers and word processors.

Labuz, Ronald. 1984. *The Interface Data Book for Word Processing/Typesetting 1984–1985.* R.R. Bowker, New York. Provides a directory of materials and equipment for interfacing word processors with typesetters. Of more use to typesetters and publishers than authors.

Lee, Marshall. 1979. *Bookmaking. The illustrated guide to design/production/editing. 2nd edition.* R.R. Bowker, New York, 485 pp. The best text on book design and production. Excellent coverage of typesetting, printing, paper, binding, planning, and budgeting. Although oriented primarily to trade books, many elements of textbook publishing are covered. A fun book to skim because of its excellent illustrations.

Lindey, Alexander. 1980. *Lindey on Entertainment, Publishing and the Arts. 2nd edition. Volume 1.* Clark Boardman Co., New York. pp. 175–244 covers book contracts. Although textbook authors may find it disquieting to discover that their activity is lumped with entertainment, unfortunately book publishing is often considered in the business world as an entertainment enterprise. This book, which should be available in all law libraries, provides practical suggestions for how the publishing contract should be written. I was shocked to learn that Jonathan Swift only earned £300 from

Gulliver's Travels, even though it was highly successful, and that Henry Fielding sold the copyright in *Tom Jones* outright for £600.

Literary Market Place. R.R. Bowker, New York. Appears annually. This book, subtitled *The Directory of American Book Publishing,* provides addresses, telephone numbers, names of lead personnel, and other useful items, for U.S. and Canadian Book Publishers, Literary Agents, Associations, Consultants, Book Producers, Editorial and Art Services, Typing Services, Translators, Photographers, Stock Photo Agencies, Book Review Journals, Book Manufacturers, and many other categories. The book also has an extensive yellow pages where the name, address, and telephone number of everyone listed in the directory is given. This book should be available in any major library.

McGraw-Hill Style Manual. A Concise Guide for Writers and Editors. McGraw-Hill Book Company, New York. 333 pp. Although there are many style manuals, this is one of the better ones for technical and scientific writing. Does not replace the Chicago Manual of Style.

National Association of College Stores. 528 E. Lorain St., Oberlin, OH 44074. This is the trade organization of college bookstores. It publishes a directory of textbook publishers which is more detailed than that published by the Association of American Publishers.

One Book/Five Ways. William Kaufmann, Inc., Los Altos, CA 94022. This book was the result of an experiment run by the American Association of University Presses in which the same manuscript was processed from original submission through the penultimate steps of production by five university presses. Complete details of the publishing procedures of each press are given, including exact copies of all contracts and forms, all financial calculations, editorial and design decisions, and marketing and sales approaches. Although the presses are noncommercial, this book provides the closest insight into the inner workings of a publisher.

Pocket Pal. A Graphics Arts Production Handbook. 12th edition. International Paper Company, 220 East 42nd St., New York, NY 10017. This widely quoted little manual provides an excellent brief introduction to the printing process.

Publishers Trade Book Annual. R.R. Bowker, New York. This 4 volume work, which appears annually, provides exact copies of publishers catalogs. Although the main emphasis is on trade books, most textbook publishers are represented because they publish both trade and textbooks. A very useful guide to who is publishing what.

Publishers Weekly. R.R. Bowker, New York. A weekly magazine of the publishing industry, available in most public libraries. Provides useful insights into the way in which the publishing industry functions today.

Rice, Stanley. 1978. *Book Design. Systematic Aspects.* R.R. Bowker, New York, 274 pp. An excellent treatment of the way in which books are designed, with good insights into how the appearance of the book can be influenced by the design. One thorough chapter deals with the design of complex textbooks.

Rice, Stanley. 1978. *Book Design. Text Format Models.* A companion to the above book, providing examples of different typographic arrangments for the various elements in a textbook. Excellent treatment of such complex textbook features as tables, footnotes, glossaries, bibliographies, mathematical displays, study materials, indexes, etc.

Strong, William S. 1984. *The Copyright Book. A Practical Guide. 2nd edition.* The MIT Press, Cambridge, MA. 223 pp. An excellent, brief, primarily nonlegal, treatment of the copyright law.

Appendix

The material that follows is supplemental and may be useful to many readers. The listing of all the members of the College Division of the Association of American Publishers will give authors an indication of the complete range of firms able to afford the AAP's stiff membership fees. Bias-free writing is essential in any textbook, and McGraw-Hill Book Company has published a little booklet on this subject, the essentials of which are reproduced here. In order to provide more details for authors interested in interfacing their computers with typesetters, material on typeset coding for word processors is reproduced from a booklet of the AAP. To supplement the discussion of the publishing contract in Chapter 9, the complete text of a model book contract is reproduced. This model contract can be used as a basis of discussion with the author's legal advisor. Although American publishers are in general highly ethical (as they must be to stay in business), an American code of ethics has apparently not been adopted. The Code of Practice published by the Publishers Association of the U.K. is reproduced in order to provide insights into the kinds of issues that often arise during author/publisher negotiations. Finally, an example is given of the kinds of marketing information that publishers require from authors in order to effectively market their books.

Members of the College Division, Association of American Publishers

ADDISON-WESLEY PUBLISHING CO.
Jacob Way
Reading, MA 01867
(617) 944-3700
TELEX: 949416

ALFRED PUBLISHING CO.
15335 Morrison Street
Sherman Oaks, CA 91403
(213) 995-8811

ANCHOR PRESS/ DOUBLEDAY & CO., INC.
245 Park Avenue
New York, NY 10167
(212) 953-4561
(800) 645-6156

ASPEN SYSTEMS CORP.
1600 Research Blvd.
Rockville, MD 20850

THE BENJAMIN/CUMMINGS PUBLISHING CO.
Editorial & Sales Offices:
2727 Sand Hill Road
Menlo Park, CA 94025
(415) 854-6020
Telex: 94 9416

BNA BOOKS
1231 25th Street, N.W.
Washington, D.C. 20037
(202) 452-4276
Telex: 892692

BOBBS-MERRILL EDUCATIONAL PUBLISHING
4300 West 62nd Street
P. O. Box 7080
Indianapolis, IN 46206
(317) 298-5686
Telex: 2-7343

WM. C. BROWN PUBLISHERS
2460 Kerper Blvd.
Dubuque, Iowa 52001
(319) 588-1451

CAMBRIDGE UNIVERSITY PRESS
32 East 57th Street
New York, NY 10022
(212) 688-8885
(800) 431-1580
Telex: 424689

CBI PUBLISHING CO.
A Division of Van Nostrand Reinhold
7625 Empire Drive
Florence, Kentucky 41042

CBS EDUCATIONAL AND PROFESSIONAL PUBLISHING
383 Madison Avenue
New York, NY 10017
(212) 872-2000

Imprints: Holt, Saunders College, Dryden, Praeger

COMPUTER SCIENCE PRESS, INC.
11 Taft Court
Rockville, MD 20850
(301) 251-9050

CREATIVE COMPUTING PRESS
One Park Avenue
New York, NY 10016

HARLAN DAVIDSON, INC.
3110 North Arlington Heights Road
Arlington Heights, IL 60004
(312) 253-9720

F.A. DAVIS COMPANY
1915 Arch Street
Philadelphia, PA 19103
(215) 568-2270
Telex: 83-4837

DELLEN PUBLISHING CORPORATION
3600 Pruneridge Avenue, Suite 340
Santa Clara, CA 95051
(408) 246-4215
(408) 245-5001

DRYDEN PRESS
(see CBS Educational and Professional Publishing)

ELS PUBLICATIONS
(English Language Services)
5761 Buckingham Parkway
Culver City, CA 90230
(213) 642-0994
TWX: 910 328 7211 ELS CULV

ELSEVIER SCIENCE PUBLISHING CO., INC.
52 Vanderbilt Avenue
New York, NY 10017
(212) 867-9040

ENSLOW PUBLISHERS
Bloy St. & Ramsey Ave.
P.O. Box 777
Hillside, NJ 07205
(201) 964-4116

FOLLETT COLLEGE BOOK COMPANY
1000 West Washington Blvd.
Chicago, IL 60607
(312) 666-4301
Telex: 25-3225
DDD: 312 666-5864

W.H. FREEMAN & COMPANY
4419 West 1980 South
Salt Lake City, Utah 84104
(801) 973-4660
Telex 453108

GINN CUSTOM PUBLISHING
(Division of Ginn and Company Publishers)
191 Spring Street
Lexington, MA 02173
(617) 863-7500

HACKETT PUBLISHING CO., INC.
832 Pierson St.
P.O. Box 44937
Indianapolis, IN 46204
(317) 635-9250

HAMMOND INCORPORATED
515 Valley Street
Maplewood, NJ 07040

HARCOURT BRACE JOVANOVICH, INC.
1250 Sixth Avenue
San Diego, CA 92101
(619) 231-6616

HARPER & ROW PUBLISHERS
10 East 53rd Street
New York, NY 10022
(212) 207-7000

HARVARD UNIVERSITY PRESS
79 Garden Street
Cambridge, MA 02138
(617) 495-2600
Telex 921484

HAYDEN BOOK CO.
50 Essex Street
Rochelle Park, NJ 07662
(201) 843-0550

D.C. HEATH & CO.
125 Spring Street
Lexington, MA 02173
(617) 862-6650

HOLT, RINEHART & WINSTON
(see CBS Educational and Professional Publishing)

HOUGHTON MIFFLIN
One Beacon Street
Boston, MA 02107
(617) 725-5000

HUMAN SCIENCES PRESS, INC.
72 Fifth Avenue
New York, NY 10011
(212) 243-6000

JAMESTOWN PUBLISHERS
Post Office Box 6743
Providence, RI 02940
(401) 351-1915

JONES AND BARTLETT PUBLISHERS, INC.
(Formerly Science Books International, Publishers)
20 Park Plaza
Boston, Mass. 02116
(617) 426-5246

LAROUSSE & CO., INC.
572 Fifth Avenue
New York, NY 10036
(212) 575-9515
Telex 12335

LEA & FEBIGER
600 Washington Square
Philadelphia, PA 19106
(215) 922-1330
Telex: 7106 701972

J.B. LIPPINCOTT CO.
The Health Professions
Publisher of Harper & Row
East Washington Square
Philadelphia, PA 19105
(215) 574-4200

LITTLE, BROWN & CO.
34 Beacon Street
Boston, MA 02106
(617) 227-0730
Telex: 92-3450

LONGMAN, INC.
19 West 44th Street
New York, NY 10036
(212) 764-3950

LOVE PUBLISHING CO.
1777 South Bellaire Street
Denver, CO 80222
(303) 757-2579

McGRAW-HILL BOOK CO.
1221 Avenue of Americas
New York, NY 10020
(212) 512-2000

McKNIGHT PUBLISHING CO.
P.O. Box 2854
Bloomington, IL 61701
(309) 663-1341

MACMILLAN PUBLISHING CO., INC.
866 Third Avenue
New York, NY 10022
(212) 702-2000

MARINER PUBLISHING CO., INC.
10927 N. Dale Mabry
Tampa, FL 33618
(813) 962-8136

MATRIX PUBLISHERS, INC.
8285 SW Nimbus, Suite 151
Beaverton, OR 97005
(503) 646-2713
(800) 547-1842

MAYFIELD PUBLISHING COMPANY
285 Hamilton Avenue
Palo Alto, CA 94301
(415) 326-1640

MERRIAM-WEBSTER INC.
47 Federal Street
P.O. Box 281
Springfield, MA 01102
(413) 734-3134

CHARLES E. MERRILL PUBLISHING COMPANY
1300 Alum Creek Drive
Columbus, OH 43216
(614) 258-8441
(800) 848-6205

MITCHELL PUBLISHING, INC.
915 River Street
Santa Cruz, CA 95060
(408) 425-3851

C.V. MOSBY CO.
11830 Westline Industrial Drive
St. Louis, MO 63141
(314) 872-8370
(800) 325-4177

NATIONAL LEARNING CORPORATION
212 Michael Drive
Syosset, NY 11791
(516) 921-8888

W.W. NORTON & COMPANY, INC.
500 Fifth Avenue
New York, NY 10110
(212) 354-5500

OXFORD UNIVERSITY PRESS
200 Madison Avenue
New York, NY 10016
(212) 679-7300
Telex: 130479

PENGUIN BOOKS
40 West 23rd St.
New York, NY 10010
(212) 807-7300
Cable: VikPress

PENNWELL BOOKS
P.O. Box 21288
Tulsa, OK 74121
(918) 663-4220

PERGAMON PRESS, INC.
Maxwell House
Fairview Park
Elmsford, NY 10523
(914) 592-7700

PETERSON'S GUIDES, INC.
166 Bunn Drive
P.O. Box 2123
Princeton, NJ 08540
(609) 924-5338

PRAEGER PUBLISHERS

(see CBS Educational and Professional Publishing)

PRENTICE-HALL, INC.
Englewood Cliffs, NJ 07632
(201) 592-2000

PWS PUBLISHERS
(Division of Wadsworth, Inc.)
Statler Office Building
20 Providence Street
Boston, MA 02116
(617) 482-2344
Telex: 940-491

PRINDLE, WEBER & SCHMIDT
(617) 482-2344

DUXBURY PRESS
(617) 482-2344

WILLARD GRANT PRESS
(617) 482-2344

GEORGE PHILIP RAINTREE
205 W. Highland Ave.
Milwaukee, WI 53203
(414) 273-0873

RANDOM HOUSE/ALFRED A. KNOPF
201 East 50th Street
New York, NY 10022
(212) PL 1-2600

REAL ESTATE EDUCATION CO.
500 North Dearborn Street
Chicago, IL 60610
(312) 836-4400

REGENTS PUBLISHING CO., INC.
2 Park Ave.
New York, NY 10016
(212) 889-2780

RESTON PUBLISHING CO., INC.
11480 Sunset Hills Road
Reston, VA 22090
(703) 437-8900

ST. MARTIN'S PRESS
175 Fifth Avenue
New York, NY 10010
(212) 674-5151
(800) 221-7945
TWX: 710-581-6459

SALEM SOFTBACKS
Penthouse
4529 Angeles Crest Hwy.
La Canada, CA 91011
(213) 790-8120

SAUNDERS COLLEGE

(see CBS Educational and Professional Publishing)

W.B. SAUNDERS CO.
West Washington Square
Philadelphia, PA 19105
(215) 574-4700
Health Science Titles only; for College Titles, see Holt listing.

SCHOCKEN BOOKS
200 Madison Avenue
New York, NY 10016
(212) 685-6500

SCHOLASTIC TESTING SERVICE, INC.
P.O. Box 1056
480 Meyer Road
Bensenville, IL 60106
(312) 766-7150

SCIENCE BOOKS INTERNATIONAL

(see Jones & Bartlett)

SCIENCE RESEARCH ASSOCIATES, INC.
College and Vocational Studies Division
155 N. Wacker Drive
P.O. Box 5380
Chicago, IL 60680-5380

SCOTT, FORESMAN & CO.
1900 East Lake Avenue
Glenview, IL 60025
(312) 729-3000

SCRIBNER BOOK COMPANIES INC.
597 Fifth Avenue
New York, NY 10017
(212) 486-2700

SINAUER ASSOCIATES INC., PUBLISHERS
Sunderland, MA 01375
(413) 665-3722

SOUTH-WESTERN PUBLISHING
5101 Madison Road
Cincinnati, OH 45227
(513) 271-8811
(800) 543-1985

UNIVERSITY PRESS OF AMERICA
4720 Boston Way
Lanham, MD 20706
(301) 459-3366

WADSWORTH, INC.
7625 Empire Dr.
Florence, KY 41024
(606) 525-2230
(800) 354-9706
Telex: 214239

BRETON PUBLISHERS
BROOKS/COLE
KENT PUBLISHING
LIFETIME LEARNING PUBLICATIONS
WADSWORTH PUBLISHING COMPANY

FRANKLIN WATTS COLLEGE DIVISION
387 Park Ave. So.
New York, NY 10016
(212) 686-7070

WEST PUBLISHING CO.
50 W. Kellogg Blvd.
P.O. Box 3526
St. Paul, MN 55165
(612) 228-2500

JOHN WILEY & SONS, INC.
605 Third Avenue
New York, NY 10158
(212) 850-6000
Telex: 12-7063

WORTH PUBLISHERS, INC.
444 Park Avenue South
New York, NY 10016
(212) 689-9630

Courtesy of the Association of American Publishers

WRITING BIAS-FREE TEXT

Outlined below are specific techniques for writing bias-free text. A careful and accurate choice of words often provides a ready solution.

Using Parallel Language

Parallel language should be used for women and men. Avoid the subtle stereotyping of women by roles. Women should be referred to as *ladies* only when men are being referred to as *gentlemen*. Similarly, women should be called *wives* and *mothers* only when men are referred to as *husbands* and *fathers*. Use parallel adjectives to describe parallel situations.

NO	*YES*
men and *ladies*	*men* and *women* *or* *ladies* and *gentlemen*
The *man* and *wife* are both economists.	The *husband* and *wife* are both economists.
The *girls'* gym is on the right; the *men's* gym is on the left.	The *women's* gym is on the right; the *men's* gym is on the left. *or* The *girls'* gym is on the right; the *boys'* is on the left.

Names

Women should not be identified in terms of their roles as wife, mother, sister, or daughter unless these roles are specifically at issue. Nor should their marital status be indicated unless the form is paired with similar references to men.

Men's and women's names should be treated comparably.

NO	*YES*
Albert Einstein and Mrs. Mead	Dr. Einstein and Dr. Mead *or* Albert Einstein and Margaret Mead
Einstein and Mrs. Mead	Einstein and Mead

Courtesy of the McGraw-Hill Book Company

Unnecessary reference to or emphasis on a woman's marital status should be avoided. A woman should be referred to by the name which she prefers, whether that name is her birth name or her name by marriage.

Titles

The same nomenclature should be used for the same job or position whether it is held by a man or a woman. Whenever possible, terms should be used which include both sexes.

NO	*YES*
Francis MacIntyre, *chairman* of the policy committee, and Alison Copely, *chairwoman* of the executive committee	Francis MacIntyre, *chair* of the policy committee, and Alison Copely, *chair* of the executive committee *or* Francis MacIntyre, *chairperson* of the policy committee, and Alison Copely, *chairperson* of the executive committee

Gender-specific pronouns should not be used to refer to workers in particular occupations, or to individuals engaged in particular activities, on the assumption that such individuals are always (or usually) female or male. Instead, use a plural form, or use *he or she* or *she or he*.

NO	*YES*
the consumer or shopper . . . she	consumers or shoppers . . . they
the assistant . . . she	the assistants . . . they
the breadwinner . . . his earnings	the breadwinner . . . his or her earnings *or* breadwinners . . . their earnings

Including Women as Participants

Women should be spoken of as participants in the world in their own right, not as appendages of men. Terms like *pioneer*, *farmer*, and *settler* should clearly include females as well as males.

NO	*YES*
Pioneers moved west, taking their wives and children with them.	Pioneer families moved west.

Women should be recognized for their own achievements.

NO	*YES*
Mrs. Paul Sager, whose husband is president of Software Enterprises and has been instrumental in the Save the Parks campaign, was granted tenure today by the English Department of City University.	Helen Sager was granted tenure today by the English Department of City University.

Women should not be portrayed as needing permission from men in order to function in the world or exercise their rights (except, of course, for historical or factual accuracy).

NO	*YES*
George Weiss allows his wife, Ruth, to work part time.	Ruth Weiss works part time.

Intelligent, daring, and innovative women should be included in both historical and fictional contexts and should be treated with respect.

Describing Women and Men

Women and men should be treated with the same respect, dignity, and seriousness. Neither sex should be trivialized or stereotyped. Women should not be described in terms of physical attributes when men are being described in terms of mental attributes or professional status. Instead, both sexes should be dealt with in the same terms. References to a man's or a woman's appearance, charm, or intuition should be avoided when irrelevant.

NO	*YES*
Mike Rivera is a respected law officer, and his wife, Annette, is a striking blond.	The Riveras are highly respected in their fields. Annette is a noted musician, and Mike is a successful law officer. *or* The Riveras are an attractive couple.

Language which assumes that the subject being written about is always either male or female should be avoided.

NO	*YES*
Doctors often neglect their wives and children.	Doctors often neglect their families.
The staff and their wives were invited.	The staff and their spouses were invited.

NO	*YES*
Give your wife a copy of this diet so that she can prepare your meals accordingly.	You and your spouse should go over this diet together so that you can plan meals at home that will be mutually satisfying and still enable you to follow the diet.
Make sure that your cap, dress, hose, and shoes are clean and neat, because an unkempt uniform can be offensive to patients.	Make sure that your clothing is clean and neat, because an unkempt uniform can be offensive to patients.

In the overall context, women should not be represented merely as sex objects or portrayed as weak, helpless, or hysterical; they should not be depicted as absurd or ridiculous, nor should their concerns be represented as trivial, humorous, or unimportant. A patronizing tone should be avoided.

Although many of the inappropriate terms listed on the next page are being used less and less, they still appear occasionally and should be avoided.

NO	*YES*
the *fair sex*, the *weaker* sex	*women* or *females*
the *girls* or the *ladies* (in reference to adult females)	the *women*
I'll ask my *girl* to check that.	I'll ask my *assistant* (or *secretary*) to check that.
lady used as a modifier, as in *lady lawyer*	*lawyer*
female-gender word forms, such as *authoress*, *poetess*, *Jewess*	*author*, *poet*, *Jew*
actor/actress	These terms and others like them (such as *waiter/waitress*) are widely accepted, but some women prefer the term *actor*. Follow individual preference if known.
diminutive word forms such as *suffragette, usherette, aviatrix*	*suffragist*, *usher*, *aviator*
libber	*feminist*
coed (as a noun)	*student*
housewife	*homemaker*
career girl or *career woman*	Name the person's profession.
cleaning lady, *cleaning woman*	*housekeeper*, *house cleaner*, or *office cleaner*

Writers should avoid showing a "gee-whiz" attitude toward women who perform competently.

NO	*YES*
Although a woman, she ran the business efficiently.	She ran the business efficiently.

Similarly, descriptions of men, especially in the context of home and personal life, should not caricature or stereotype them. Men should not be shown as dependent upon women for advice on what to wear, what to eat, and so on. Nor should they be characterized as inept in household maintenance or child care. Such expressions as *henpecked husband* and *boys' night out* should be avoided.

Gender-specific terms should be replaced whenever possible by terms that can include members of either sex.

NO	*YES*
Today, businessmen are doing billions of dollars of business.	Today, companies are doing . . . *or* Today, businesspeople are doing . . .
John Taylor is a salesman for a construction business, and his wife is a saleswoman for a word-processing firm.	The Taylors are both sales agents. John sells lumber for a building supply company, and Gloria is a sales representative for a word-processing firm.

Avoiding "Man" Words

In references to humanity at large, gender-specific terms should be avoided whenever possible. The word *man,* long used to denote all of humanity, perpetuates bias. The following alternatives are recommended.

NO	*YES*
mankind, man	*humanity, human beings, human race, people, men and women*
man's achievements	*human achievements, society's achievements*
manpower	*human power, human energy, workers, work force, personnel*
manhood	*adulthood, manhood or womanhood*
man-made	*artificial, synthetic, manufactured, constructed, of human origin*
the *average man*	the *average person*
This phenomenon has been observed in *men* and other mammals.	This phenomenon has been observed in *humans* and other mammals.
There are only six crews available to *man* the trucks.	There are only six crews available to *operate* the trucks.

Choosing Figures of Speech

Figures of speech which are, or seem to be, sexist should be avoided. Expressions like "man in the street" are best reworded entirely rather than replaced with "man or woman in the street."

NO	*YES*
old wives' tales	*superstitions, folk wisdom*
man in the street	*average citizen, average person, average American*
sob sister	*exploitative journalist, do-gooder, sentimentalist*
right-hand man	*closest associate*
straw man	*unreal issue, misrepresentation*
yes-man	*sycophant*

Occasionally, a historical context makes such an expression appropriate: *no-man's land*, for example, in the context of World War I. As a rule, however, rewording avoids bias and improves clarity.

Note that descriptive terms deriving from historical or fictional personages are considered bias-free insofar as they can be applied to persons of either sex: for example, a *Pollyanna*, a *Cassandra*, a *quisling*.

Finally, some expressions, "tomboy," for example, embody intrinsically biased ideas. In such expressions, the concept itself—not just the choice of words—should be questioned.

Using Pronouns

The English language lacks a nongender-specific singular personal pronoun. Although masculine pronouns have generally been used for reference to a hypothetical person or to humanity in general (in such constructions as "anyone . . . he" and "each child opened his book"), the following alternatives are recommended as preferable.

1 Recast to eliminate unnecessary gender-specific pronouns.

NO	*YES*
When a mechanic is checking the brakes, he must observe several precautions.	When checking the brakes, a mechanic must observe several precautions.

2 Use a plural form of the pronoun—a very simple solution and often the best.

	When mechanics check the brakes, they must observe several precautions.

3 Recast in the passive voice, making sure there is no ambiguity.

When a mechanic is checking the brakes, several precautions must be observed.

4 Use *one* or *we*.

When checking the brakes, we must observe several precautions.

5 Use a relative clause.

A mechanic who is checking the brakes must observe several precautions.

6 Recast to substitute the antecedent for the pronoun.

NO	*YES*
The committee must consider the character of the applicant and his financial responsibilities.	The committee must consider both the character and the financial responsibilities of the applicant.

7 Use *he or she*, *him or her*, or *his or her*. (These constructions have passed easily into wide use.)

The committee must consider the character of the applicant and his or her financial responsibilities.

8 Alternate male and female expressions and examples.

NO	*YES*
I've often heard supervisors say "He's not the right man for the job," or "He lacks the qualifications for success."	I've often heard supervisors say, "She's not the right person for the job," or "He lacks the qualifications for success."

Generic coding for typesetting from computer files

If you have made the decision to prepare your manuscript on a WP system, there may be the possibility of a typesetting **interface.** WP systems and computer typesetting machines store text information in basically the same way. In an interface, text prepared on one machine is translated electronically (or **converted**) by means of special software into information that is meaningful to the other.

Theoretically, any WP manuscript should be fodder for any typesetting machine. In reality, there are many complexities that may make conversion not feasible for specific systems. The specifics of translation software vary with hardware, WP software, and even versions of software. If you get a new, improved version of your software halfway through your book, it just might mean that conversion would require two different sets of translation software – and one of those sets may have to be specially developed for the task.

If your disks *can* be converted economically, it may not be necessary for the typesetter to rekeyboard your manuscript if you submit disk copies of your final manuscript along with the paper version. To facilitate this conversion, your publisher may ask you to insert special typesetting command codes in the manuscript as you type.

Before you begin keyboarding, contact your publisher. Your editor will put you in touch with in-house specialists who can discuss conversion typesetting with you and answer your questions. They will have many specific questions of their own about the hardware and software you are going to use, and they may wish to test some disks prepared on your system. The feasibility of conversion has to be established before deciding on coding.

Assuming that an interface with the typesetter is possible, you and your publisher will need to agree on whether you should code your manuscript and on the exact procedures you should follow.

SPECIAL KEYBOARDING INSTRUCTIONS

In consultation with your publisher, you will receive instructions for keyboarding your manuscript in a consistent way that will be most useful for the typesetting process. You may also receive detailed instructions for "generic coding," which will be described in the next section of this chapter.

One key to successful interfacing is consistency in the electronic medium provided to the publisher and eventually to the typesetter. The rule is: *Always do the same thing the same way throughout the manuscript.* Whatever list of "dos and don'ts" is followed, the important thing is to follow

Courtesy of the Association of American Publishers

it consistently all the way through. Here is a list of conventions similar to those your publisher may give you. They are typical instructions; conventions differ.

1. Type all copy flush left. Do not center any copy. Do not use the spacebar or any other word processing function code to achieve centering. Use the spacebar only for word separation. Some word processors store tabs and centered material in a way that is difficult for typesetting machines to work with.
2. Do not hyphenate a word at the end of a line. Hyphenate only compound words (such as *mother-in-law*); use only hyphens that are to appear in print (another example, *him- or herself*). It may be difficult for the typesetting machine to distinguish between optional hyphens at the ends of lines and required hyphens like the ones in the examples.
3. Do not use underscoring. Where italic type is desired, it can be indicated by a code. This will be explained in the next section. The word processor may store underscoring as a series of separate characters rather than associating it with specific text. The typesetter may then not be able to translate the underscoring into a change to italic typeface.
4. Never key a carriage return at the end of a line within a paragraph. Always press the spacebar when reaching the end of a line on your screen, as if the whole paragraph were only one long line. The word processor will "wrap" the lines around on your screen as if you had thrown the carriage return. A carriage return, on the other hand, may be interpreted by the typesetter as a command to begin a new paragraph.
5. Always type mnemonic codes in lower case. Always keying them upper case would be acceptable, but would require additional keystrokes, which translates to more time, effort and errors.
6. Always precede a code with a space, except when a code begins the line.
7. Letters of the alphabet should not be used to represent special characters. Do not, for example, use an "o" for a bullet. Special character keys should not be used to represent more than one legitimate print character. In other words, any character typed must always represent the same thing and only that.
8. In keyboarding quotation marks, use *two double* quotes for the opening quote and *two single* quotes for the closing quote. For example, ""*Now is the time.*''
9. Always key "1" for the number "one," and key "l" for the letter "ell." Do not use the key for the letter to represent the number, as you might on a standard typewriter.
10. Highlighted phrases within the text, such as those which might be in italics or bold, should be coded. For example, in the following phrase,

the word "all" is coded to set in italics: Now is the time for [i]all [r]good people. (The [i] begins setting in italics; the [r] recommences setting in roman.) Underscoring and boldface codes used in some WP systems – but not all – can be converted to italic and boldface type, respectively.

11. Type all footnotes at the end of each chapter. Insert the following code within the text where the footnote reference number is to appear: [fnref] 1, 2, 3, etc.

GENERIC CODING

Generic coding simply identifies each part of the manuscript that shares the same typesetting specifications. These parts are called the *elements* of the manuscript. Some examples of possible manuscript elements are: chapter title, chapter number, text, number-one head, number-two head, numbered list, and footnote. Most generic codes are mnemonic, and they are often referred to as **mnemonic codes.** Some examples of mnemonic codes are: "ct" for "chapter title," "cn" for "chapter number," "te" for "text," "h1" for "number-one head," "h2" for "number-two head," "nl" for "numbered list," and "fn" for "footnote."

SAMPLE MNEMONIC CODES

[pn]	part number
[pt]	part title
[pst]	part subtitle
[ptx]	part text
[cn]	chapter number
[ct]	chapter title
[cst]	chapter subtitle
[h1]	first-level head
[h2]	second-level head
[h3]	third-level head
[h4]	fourth-level head
.	
.	
.	
[tx]	text
[nl]	numbered list
[unl]	unnumbered list
[li]	list item
[ext]	extract
[fn]	footnote
[fnref]	footnote reference
[fgn]	figure number

[fgt]	**figure title**
[sn]	**source note**
[tt]	**table title**
[tn]	**table number**
[tch]	**table column heads**
[tb]	**table body**
[tfn]	**table footnote**
[gl]	**glossary**
[ref]	**reference**
[r]	**Roman, regular typeface**
[i]	**italic typeface**
[b]	**bold typeface**
[bi]	**bold italic typeface**
[help]	**Use when you do not know how to identify an element. Simply type this word between your chosen delimiters.**

If each element in a manuscript is identified by a different generic code, the typesetter can then write a program for the computer that will convert the generic codes into typesetting specifications for each element. Because generic codes serve only to identify the parts of the manuscript, they can be keyed into the manuscript before the book is designed. Their usefulness is not affected by changes in type specifications.

To distinguish code characters from text characters, code characters must be identified by **delimiters,** specific characters or character sequences that never appear elsewhere in the text. Codes may be surrounded with brackets [ ], braces { }, angle brackets <>, or some other set of characters that are not otherwise used throughout the entire manuscript. Parentheses ( ) and slashes / / are not good delimiters because they are usually needed within the body of the text itself. Codes may also be set off with an unlikely sequence of characters such as colon period, :. or a dollar mark followed by a letter, $x.

Once you and your publisher decide on a delimiter, it must be used consistently throughout the manuscript. Delimiters tell the typesetting machine where to find code information. A complete code is made up of both delimiter and code information, and must have both parts to be functional.

SAMPLE MANUSCRIPT

The following pages show a manuscript with generic codes, followed by the same manuscript after being set in type.

[cn]3

[ct]Historical Foundations of Systematic Counseling

[h1]OVERVIEW

[tx]Since the systems approach is a recent and relatively unknown innovation in the behavioral sciences, we think it is important to provide you with information about the historical foundations of Systematic Counseling. A few of the major events and people who played a part in the development of the systems approach are described in this chapter. The information contained in the previous chapter and this one should provide you with an adequate understanding of the philosophical and historical foundations of Systematic Counseling. This knowledge will also assist you to use the model of Systematic Counseling more effectively.

[h1]INTRODUCTION

[h2]Organization

[tx]This chapter consists of two major sections. A general discussion of some of the major historical developments of the scientific systems approach is presented in the fist section. The influence that this approach is having on several disciplines is also emphasized. The second section describes the brief history of the application of the systems approach to problems presented by counseling and guidance services.

[h2]Purpose

[tx]The major purpose of this chapter is to provide the historical foundations and context of Systematic Counseling.

[h2]Objectives

[tx]After studying this chapter you should be prepared to do the following:

[nl] [li]1. State the purpose of general systems theory.

[li]2. Name two important functions served by general systems theory.

[li]3. Name at least three disciplines in which developmental work in general systems theory has been done.

[li]4. State the functions of an operations research team.

[li]5. Describe at least two specific situations in which operations research has been conducted.

[li]6. Name the three events which occurred between 1946 and 1950 that enabled scientists to conceptualize a systems approach to their work.

[li]7. Explain the importance of SAGE and the early NASA projects such as Explorer I and Mercury in the development of systems engineering.

[li]8. Name the person who apparently was the first to describe the use of the systems approach in counseling and state why he and his associates were preoccupied with man-machine systems.

[li]9. List at least five uses of the systems of approach in guidance and counseling that do not necessarily include man-machine systems.

[h1]THE DEVELOPMENT OF THE SYSTEMS APPROACH

[tx]The history of the systems approach, although brief in terms of time span, is complicated to trace because of the vast number of

related theories, models, and projects generated by this approach in a host of disciplines and technical fields. The explosion of knowledge in this area has been so great that several volumes would be needed to adequately trace all of the developments. Therefore, we have arbitrarily concentrated on three areas in order to provide a brief overview of the evolution of the systems approach. These are:

[nl] [li]1. General systems theory
[li]2. Operations research
[li]3. Systems engineering

[tx]The systems approach, as it is known today, has been shaped by the developments in each of these areas. Concepts from each have been influential in the development and design of Systematic Counseling.

[h2]General Systems Theory

[tx]Bertalanffy, a biologist, is usually credited (DeGreene, 1970, p. 14; Khailov, 1968, p. 47; Kremyanskiy, 1968, p. 77) with introducing the idea of general systems theory in 1937. He noted the structural similarity of mathematical expressions and models used in biological, behavioral, and social sciences (Bertalanffy, 1968, p. 13). He felt that these similarities or isomorphisms were being ignored by science. Furthermore, he believed it was possible to develop a higher generality for science (a general theory) from these similarities which could serve as a foundation for the specific theories already developed by the various branches of science. General systems theory would thus, ideally, bring about the eventual unification of the various sciences.

The purpose of general systems theory, then, is to identify the common ways in which the components of different systems are organized or interrelated. Principles or concepts that apply in a universal way to several specific theories could then be investigated. Such a general theory would, of course, enhance communication across the various sciences and provide a holistic rather than a reductionistic approach to scientific knowledge. This purpose is concisely expressed in the following description of the major functions of the Society for General Systems Research.

> [ext]Major functions are to: (1) investigate the isomorphy of concepts, laws, and models in various fields, and to help in useful transfers from one field to another; (2) encourage the development of adequate theoretical models in the fields which lack them; (3) minimize the duplication of theoretical effort in different fields; (4) promote the unity of science through improving communication among specialists (Bertalanffy, 1968, p. 15).

[tx]The requirements that a general theory of systems should satisfy have been pointed out by Mesarovic. These are:

[nl] [li]1. The general theory should be general enough to encompass different types of already existing specific theories. It should, therefore, be sufficiently abstract so that its terms and concepts are relevant to specialized theories. Clearly, the more abstract statements have a broader content but, at the same time, they carry less information regarding the behavior of any particular system. The general concepts must emphasize the common features of all the systems considered yet neglect the specific aspects of the behavior of any particular

system. The real challenge in developing a general theory is, therefore, to find the proper level of abstraction. The concepts must have wide application, while the conclusions which they lead to must provide sufficient information for proper understanding of the particular class of phenomena under consideration.

[li]2. The general theory has to have a scientific character in the sense that its concepts and terms must be uniquely defined within the proper context. If the general theory is to be of any help in solving scientific and engineering problems, it must not rely on vague, ill-defined, almost poetic analogies. The basis for the general theory must be solid so that its conclusions have practical meaning for real systems....(1964, p.3)

[tx]Mesarovic's statements suggest that a general systems theory would serve an integrative function for the amalgamation of the concepts and research of several theories which fall within a given class of phenomena. The need for such an overarching general theory has been discussed by Rapoport (1968, p. xxi), who points out how the trend toward narrow specialization in science has posed a threat of an ""avalanche of 'findings' which in their totality no more adds up to knowledge, let alone wisdom, than a pile of bricks adds up to a cathedral.' '

Any serious student of psychology realizes the appropriateness of Rapoport's comment when he attempts to read and synthesize the voluminous amount of literature produced by this one discipline. The applied behavioral scientist, and especially the practitioner, desperately needs a general theory of human behavior which has the capability for organizing and relating current and future knowledge within an overall framework. A general systems theory of human behavior could serve to integrate and synthesize the knowledge produced by the disciplines of anthropology, economics, history, human ecology, human physiology, philosophy, psychology, and sociology. At the present time it is impossible for a counselor, as one practitioner of an applied behavioral science, to utilize all of the potentially useful information from these disciplines in attempting to understand and assist clients.

As several writers have indicated (Bertalanffy, 1969, p. 36; Gray and Rizzo, 1969, p. 26; and DeGreene, 1970, p. 25), we are only at the beginning stage in developing a general systems theory. However, a great deal of developmental work has been done in several disciplines. We have already mentioned the work of Bertalanffy, whose work has been influential in many fields and especially in biology and psychiatry. The following examples will illustrate the scope and influence of general systems theory in several disciplines.

[h3]Sociology. [r]Buckley (1968) views society or the sociocultural system as a complex adaptive system which is open and subject to change. He has outlined a model for such a system which he believes can unite some of the more recent sociological and social psychological theories.

[h3]Archaeology. [r]Clarke (1968) has created a general model for the organization and relation of archaeological activities. His model starts with an archaeological segment of the real world. This sample is subjected to experimentally controlled contextual and specific observations from which an hypothesis or model is constructed. The model or hypothesis is then tested against the latest observation samples for goodness of fit. Propositions about the real world

based on the archaeological segment are then synthesized.

[h3]Human Factors Engineering and Psychology. [r]The works of DeGreene (1970) and Gagne (1962) contain several models and concepts derived from systems theory which are applied in the fields of human factors engineering and psychology. In the terminology of that period, such topics as man-computer interrelationships, man-man and man-machine communications, concepts of training, motivation, and job performance are considered.

[h3]Cognitive Activity. [r]Laszlo (1969) has endeavored to develop a scientific theory of the mind based on the concepts of general systems theory. His intent is to construct a map of the basic structure of cognitive activity. He states that the exploratory phases of his work suggest that a basic regulatory structure underlies the phenomena of the mind and of human interrelations. This regulatory structure is believed to be similar to the servomechanisms used by cyberneticians.[fnref]1

While the above examples do not represent a comprehensive survey of the total field, they do demonstrate that general systems theory has had a significant impact on a number of scientific disciplines in a relatively short period of time. In addition, general systems theory was only one aspect of the change in scientific thought that followed World War II. A number of closely related theories and concepts, although with different emphases and terminology, were in the process of development of approximately the same time.

[fnref]1 [fn]A complete and technical discussion of general systems theory as it has been utilized by the theoreticians mentioned in this section may be found by consulting the references section of this book.

3 Historical Foundations of Systematic Counseling

OVERVIEW

Since the systems approach is a recent and relatively unknown innovation in the behavioral sciences, we think it is important to provide you with information about the historical foundations of Systematic Counseling. A few of the major events and people who played a part in the development of the systems approach are described in this chapter. The information contained in the previous chapter and this one should provide you with an adequate understanding of the philosophical and historical foundations of Systematic Counseling. This knowledge will also assist you to use the model of Systematic Counseling more effectively.

INTRODUCTION

Organization

This chapter consists of two major sections. A general discussion of some of the major historical developments of the scientific systems approach is presented in the first section. The influence that this approach is having on several disciplines is also emphasized. The second section describes the brief history of the application of the systems approach to problems presented by counseling and guidance services.

Purpose

The major purpose of this chapter is to provide the historical foundations and context of Systematic Counseling.

Objectives

After studying this chapter you should be prepared to do the following:

1. State the purpose of general systems theory.
2. Name two important functions served by general systems theory.
3. Name at least three disciplines in which developmental work in general systems theory has been done.
4. State the functions of an operations research team.
5. Describe at least two specific situations in which operations research has been conducted.
6. Name the three events which occurred between 1946 and 1950 that enabled scientists to conceptualize a systems approach to their work.
7. Explain the importance of SAGE and the early NASA projects such as Explorer I and Mercury in the development of systems engineering.
8. Name the person who apparently was the first to describe the use of the systems approach in counseling and state why he and his associates were preoccupied with man-machine systems.
9. List at least five uses of the systems approach in guidance and counseling that do not necessarily include man-machine systems.

THE DEVELOPMENT OF THE SYSTEMS APPROACH

The history of the systems approach, although brief in terms of time span, is complicated to trace because of the vast number of related theories, models, and projects generated by this approach in a host of disciplines and technical fields. The explosion of knowledge in this area has been so great that several volumes would be needed to adequately trace all of the developments. Therefore, we have arbitrarily concentrated on three areas in order to provide a brief overview of the evolution of the systems approach. These are:

1. General systems theory
2. Operations research
3. Systems engineering

The systems approach, as it is known today, has been shaped by the developments in each of these areas. Concepts from each have been influential in the development and design of Systematic Counseling.

34 Historical Foundations of Systematic Counseling

General Systems Theory

Bertalanffy, a biologist, is usually credited (DeGreene, 1970, p. 14; Khailov, 1968, p. 47; Kremyanskiy, 1968, p. 77) with introducing the idea of general systems theory in 1937. He noted the structural similarity of mathematical expressions and models used in biological, behavioral, and social sciences (Bertalanffy, 1968, p. 13). He felt that these similarities or isomorphisms were being ignored by science. Furthermore, he believed it was possible to develop a higher generality for science (a general theory) from these similarities which could serve as a foundation for the specific theories already developed by the various branches of science. General systems theory would thus, ideally, bring about the eventual unification of the various sciences.

The purpose of general systems theory, then, is to identify the common ways in which the components of different systems are organized or interrelated. Principles or concepts that apply in a universal way to several specific theories could then be investigated. Such a general theory would, of course, enhance communication across the various sciences and provide a holistic rather than a reductionistic approach to scientific knowledge. This purpose is concisely expressed in the following description of the major functions of the Society for General Systems Research.

> Major functions are to: (1) investigate the isomorphy of concepts, laws, and models in various fields, and to help in useful transfers from one field to another; (2) encourage the development of adequate theoretical models in the fields which lack them; (3) minimize the duplication of theoretical effort in different fields; (4) promote the unity of science through improving communication among specialists (Bertalanffy, 1968, p. 15).

The requirements that a general theory of systems should satisfy have been pointed out by Mesarović. These are:

1. The general theory should be general enough to encompass different types of already existing specific theories. It should, therefore, be sufficiently abstract so that its terms and concepts are relevant to specialized theories. Clearly, the more abstract statements have a broader content but, at the same time, they carry less information regarding the behavior of any particular system. The general concepts must emphasize the common features of all the systems considered yet neglect the specific aspects of the behavior of any particular system. The real challenge in developing a general theory is, therefore, to find the proper level of abstraction. The concepts must have wide application, while the conclusions which they lead to must

provide sufficient information for proper understanding of the particular class of phenomena under consideration.

2. The general theory has to have a scientific character in the sense that its concepts and terms must be uniquely defined within the proper context. If the general theory is to be of any help in solving scientific and engineering problems, it must not rely on vague, ill-defined, almost poetic analogies. The basis for the general theory must be solid so that its conclusions have practical meaning for real systems. . . .(1964, p. 3)

Mesarović's statements suggest that a general systems theory would serve an integrative function for the amalgamation of the concepts and research of several theories which fall within a given class of phenomena. The need for such an overarching general theory has been discussed by Rapoport (1968, p. xxi), who points out how the trend toward narrow specialization in science has posed a threat of an "avalanche of 'findings' which in their totality no more adds up to knowledge, let alone wisdom, than a pile of bricks adds up to a cathedral."

Any serious student of psychology realizes the appropriateness of Rapoport's comment when he attempts to read and synthesize the voluminous amount of literature produced by this one discipline. The applied behavioral scientist, and especially the practitioner, desperately needs a general theory of human behavior which has the capability for organizing and relating current and future knowledge within an overall framework. A general systems theory of human behavior could serve to integrate and synthesize the knowledge produced by the disciplines of anthropology, economics, history, human ecology, human physiology, philosophy, psychology, and sociology. At the present time it is impossible for a counselor, as one practitioner of an applied behavioral science, to utilize all of the potentially useful information from these disciplines in attempting to understand and assist clients.

As several writers have indicated (Bertalanffy, 1969, p. 36; Gray and Rizzo, 1969, p. 26; and DeGreene, 1970, p. 25), we are only at the beginning stage in developing a general systems theory. However, a great deal of developmental work has been done in several disciplines. We have already mentioned the work of Bertalanffy, whose work has been influential in many fields and especially in biology and psychiatry. The following examples will illustrate the scope and influence of general systems theory in several disciplines.

Sociology. Buckley (1968) views society or the sociocultural system as a complex adaptive system which is open and subject to change. He has outlined a model for such a system which he believes can unite some of the more recent sociological and social psychological theories.

Archaeology. Clarke (1968) has created a general model for the organization and relation of archaeological activities. His model starts with an archaeological segment of the real world. This sample is subjected to experimentally controlled contextual and specific observations from which an hypothesis or model is constructed. The model or hypothesis is then tested against the latest observation samples for goodness of fit. Propositions about the real world based on the archaeological segment are then synthesized.

Human Factors Engineering and Psychology. The works of DeGreene (1970) and Gagne (1962) contain several models and concepts derived from systems theory which are applied in the fields of human factors engineering and psychology. In the terminology of that period, such topics as man-computer interrelationships, man-man and man-machine communications, concepts of training, motivation, and job performance are considered.

Cognitive Activity. Laszlo (1969) has endeavored to develop a scientific theory of the mind based on the concepts of general systems theory. His intent is to construct a map of the basic structure of cognitive activity. He states that the exploratory phases of his work suggest that a basic regulatory structure underlies the phenomena of the mind and of human interrelations. This regulatory structure is believed to be similar to the servomechanisms used by cyberneticians.[1]

While the above examples do not represent a comprehensive survey of the total field, they do demonstrate that general systems theory has had a significant impact on a number of scientific disciplines in a relatively short period of time. In addition, general systems theory was only one aspect of the change in scientific thought that followed World War II. A number of closely related theories and concepts, although with different emphases and terminology, were in the process of development at approximately the same time. The more important developments are:

1. *Cybernetics*, based upon the principle of feedback or circular causal trains providing mechanisms for goal-seeking and self-controlling behavior.

[1]A complete and technical discussion of general systems theory as it has been utilized by the theoreticians mentioned in this section may be found by consulting the references section of this book.

A model publishing contract for a college textbook

TEXTBOOK PUBLISHERS INC.

AN AGREEMENT

Between____
a citizen of____
whose home address is_______________________________________
____
____
____

hereinafter called the "Author"
and Textbook Publishers Inc. 999 West River Drive, New York
hereinafter called the "Publisher"
with respect to the publication of a work tentatively titled

____

hereinafter called the "Work".

1. GRANT OF RIGHTS

The Author grants this Work to the Publisher with the exclusive right to publish and sell the Work, under its own name and under any other name, during the full term of copyright and all extensions thereof, and to copyright it in the Publisher's name or any other name in all countries of the World; also the exclusive rights listed in paragraph 4 below, with authority to dispose of said rights in all languages and in all countries.

2. DELIVERY OF MANUSCRIPT

The manuscript, to contain about

________________________ words or their equivalent, will be delivered by the Author by

____

3. ROYALTIES

The Publisher will pay the Author the following royalty, based on the Publisher's actual dollar receipts:

________ %from sales of the regular edition in the United States, its territories, the Commonwealth of Puerto Rico, and Canada

________ %from sales of the regular edition elsewhere

________ %from sales of reprint editions, defined as editions published in less expensive form than the regular edition

________ %from sales of condensations, adaptations, and other derivative works (other than revisions) published by it and from its sales and rentals of filmstrips, slides, transparencies, tapes, records, microform, and similar media.

________ %from the sales of the regular edition directly to the consumer through Publisher-owned book clubs and sales resulting from mail order campaigns and solicitation by radio and television.

4. SUBSIDIARY RIGHTS

The Publisher may permit others to publish, broadcast by radio, make recordings or mechanical renditions, enter into electronic or digital form, show by motion pictures or by television, syndicate, quote, make translations and other versions, and otherwise utilize this work, and material based on this work. The net amount of any compensation received from such use shall be divided equally between the Publisher and the Author. If the Publisher sells any stock of the Work at a price below the manufacturing costs of the book plus royalties, no royalties shall be paid.

5. PAYMENT AND ACCOUNTING

Payments to the Author shall be made in March and September of each year for the six-month period ending the prior December 31 and June 30, respectively, and shall be accompanied by an appropriate Statement of Account. If the balance due the Author for any settlement period is less than $10, the Publisher will make no accounting or payment until the next royalty period at the end of which the cumulative balance has reached $10. The Publisher may deduct from any funds due the Author, under this or any other agreement between the Author and the Publisher, any sum that the Author may owe the Publisher.

6. SUBMISSION OF MANUSCRIPT

The final manuscript shall be delivered in double-spaced typescript or its equivalent, satisfactory to the Publisher in organization, form, content, and style. If the Author fails to deliver a satisfactory manuscript on or before the due date, unless the time for submission has been extended in writing by the Publisher, the Publisher will have the right to terminate this agreement and to recover from the Author any money advanced in connection with this work. Until this agreement has been terminated and until any advanced money has been returned, the Author may not have the Work published elsewhere.

7. ITEMS FURNISHED BY AUTHOR

The Author will furnish the following items along with the manuscript: title page; preface or foreword (if any); table of contents; index; teacher's manual or key (if requested by the Publisher); and complete and final copy for all illustrative material prepared for reproduction.

8. AUTHOR'S DUTIES

The Author will read proofs, correct them in duplicate, and promptly return one set to the Publisher. The Author will be responsible for the completeness and accuracy of such corrections and will bear all costs of alteration in the proofs (other than those resulting from printer's errors) exceeding 10%of the cost of the typesetting. Author's Alterations are defined as deletions, additions, and any other revisions made by the Author to the proofs, other than to correct printer's errors. These costs will be deducted from the royalty payments due the Author.

9. AUTHOR'S WARRANTY

The Author warrants that the Work is original on the Author's part except for such excerpts and illustrations from copyrighted works that may be included with the written permission of the copyright owner. Permission to quote from copyrighted sources shall be obtained by the Author at his or her own expense (on a form approved by the Publisher for this Work), such permissions to be submitted to the Publisher with the final manuscript. The Author further warrants that the Work does not infringe any copyright, violate any property rights, contain any libelous, scandalous, or unlawful statements, contain any instructions that may cause harm or injury, or infringe upon any trademark or other right or the privacy of others. The Author will defend, indemnify, and hold harmless the Publisher against all claims, suits, costs, damages, and expenses that the Publisher may sustain by reason of any scandalous, libelous, or unlawful matter contained or alleged to be contained in the work, or any infringement or violation by the work of any copyright or property right; and until such claim or suit has been settled or withdrawn, the Publisher may withhold any sums due the Author under this agreement.

10. PUBLICATION DETAILS

The Publisher will have the right to edit the work for the original printing and for any reprinting, provided that the meaning of the text is not materially altered. The Publisher will have the right to publish the Work in suitable style as to paper, printing, and binding; to fix or alter the title and price; and to use all customary means to market the Work.

11. AUTHOR'S COPIES

Upon publication the Publisher shall furnish the Author without charge six copies of the book. The Author may purchase for personal use additional copies at a 25%discount from the lowest list price.

12. REVISIONS

The Author agrees to revise the Work if the Publisher considers it necessary in the best interests of the Work. The provisions of this agreement shall apply to each revision of the Work by the Author as though that revision were the work being published for the first time under this agreement. Should the Author be unable or unwilling to provide a revision within a reasonable time after the Publisher has requested it, or should the Author be deceased, the Publisher may have the revision prepared and charge the cost, including, without limitation, fees or royalties, against the Author's royalties, and may display in the revised work, and in advertising, the name of the person or persons who revise the work.

13. COMPETING PUBLICATIONS

The Author agrees that during the term of this agreement the Author will not agree to publish or furnish to any other publisher any work that will conflict with the sale of this Work.

14. DISCONTINUING MANUFACTURE

When the Publisher determines that the demand for this Work no longer warrants its continued manufacture, the Publisher may discontinue manufacture and destroy any or all plates, books, and sheets without liability to the Author.

15. CONSTRUCTION, HEIRS, ARBITRATION

This agreement shall be construed and interpreted according to the laws of the State of New York. It shall be binding upon the parties hereto, their heirs, successors, assigns, and personal representatives, and any references to the Author or to the Publisher shall include their heirs, assigns, successors, and personal representatives. This agreement shall not be subject to change or modification in whole or in part, except by a written document signed by the party against whom enforcement is sought. All differences or disputes arising in connection with this agreement shall be settled by arbitration in New York City in accordance with the rules of the American Arbitration Association then obtaining, and judgement upon the award may be entered in any court having jurisdiction thereof.

____

U.S. TAXPAYER IDENTIFICATION NO. AUTHOR

TEXTBOOK PUBLISHERS INC.

BY ______________________

____

TITLE

Code of Practice
Guidelines for Publishers

A constructive and co-operative relationship between authors (and the agents and representatives acting for them) and their publishers is vital to successful publishing. In the great majority of cases this relationship undoubtedly exists. Nevertheless, there can be dissatisfaction, perhaps because a title is not the success the author and publisher hoped for but also because of misunderstandings of the publishing contract, uncertainties and poor drafting, and "customs of the trade" unappreciated by the author.

The Council of the Publishers Association belives that everything possible should be done to ensure a satisfactory relationship and avoid disputes. It has therefore prepared the Code of Practice for book publishers set out below, which it recommends to members in their dealings with authors. This Code gives guidance only. It cannot deal with every variation. In general, however, failure to accept the guidance included in the Code without good reason is clearly likely to damage the standing of individual publishers and of publishing generally.

Book publishing is so varied in its scope that contracts are likely to contain many variations between, for example, different types of book with different markets, different degrees of editorial involvement by the publisher, and established or relatively new authors. Total uniformity of contract or practice is therefore impracticable. In particular, some academic, educational, reference books and works based on a variety of contributions may be subject to special considerations, though the necessity to follow the general principles of this Code remains.

1. **The publishing contract must be clear, unambiguous and comprehensive, and must be honoured in both the letter and the spirit.**

Matters which particularly need to be defined in the contract include:–

(i) a title which identifies the work or (for incomplete works) the nature and agreed length and scope of the work.

(ii) the nature of the rights conferred – the ownership of the copyright (an assignment or an exclusive licence) whether all volume rights (or part of the volume rights or more than volume rights) and the territories and languages covered.

(iii) the timescale for delivery of the manuscript and for publication.

(iv) the payments, royalties and advances (if any) to be paid, what they are in respect of and when they are due.

(v) the provisions for sub-licensing.

(vi) the responsibility for preparing the supporting materials (e.g. indexes, illustrations, etc.) in which the author holds the copyright and for obtaining permissions and paying for the supporting materials in which the copyright is held by third parties.

(vii) the termination and reversion provisions of the contract.

Should the parties subsequently agree changes to the contract, these should be recorded in writing between them.

Courtesy of the Publishers Association, 19 Bedford Square, London

2. **The author should retain ownership of the copyright, unless there are good reasons otherwise.**

An exlusive licence should be sufficient to enable the publisher to exploit and protect most works effectively. In particular fields of publishing (e.g. encyclopaedic and reference works, certain types of academic works, publishers' compilations edited from many outside contributions, some translations and works particularly vulnerable to copyright infringement because of their extensive international sale) it may be appropriate for the copyright to be vested in the publisher.

3. **The publisher should ensure that an author who is not professionally represented has a proper opportunity for explanation of the terms of the contract and the reasons for each provision.**

4. **The contract must set out reasonable and precise terms for the reversion of rights.**

When a publisher has invested in the development of an author's work on the market, and the work is a contribution to the store of literature and knowledge, and the publisher expects to market the work actively for many years, it is reasonable to acquire volume rights for the full term of the copyright, on condition that there are safeguards providing for reversion in appropriate circumstances.

The circumstances under which the grant of rights acquired by the publisher will revert to the author (e.g. fundamental breach of contract by the publisher, or when a title has been out of print or has not been available on the market for a stipulated time) should form a part of the formal contract. In addition, a reversion of particular rights that either have never been successfuly exploited by the publisher, or which are not subject to any current (or immediately anticipated) licence or edition, may, after a reasonable period from their first acquisition and after proper notice, be returned on request to the author, provided that such partial reversions do not adversely affect other retained rights (e.g. the absence of an English language edition should not affect the licensing publisher's interest in a translated edition still in print) and provided that payments made by the publisher to or on behalf of the author have been earned.

5. **The publisher must give the author a proper opportunity to share in the success of the work.**

In general, the publishing contract should seek to achieve a fair balance of reward for author and publisher. On occasion it may be appropriate, when the publisher is taking an exceptional risk in publishing a work, or the origination costs are unusually high, for the author to assist the publication of the work by accepting initially a low royalty return. In such cases, it is also appropriate for the publisher to agree that the author should share in success by, for example, agreeing that royalty rates should increase to reflect that success.

If under the contract the author receives an outright or single payment, but retains ownership of the copyright, the publisher should be prepared to share with the author any income derived from a use of the work not within the reasonable contemplation of the parties at the time of the contract.

6. **The publisher must handle manuscripts promptly, and keep the author informed of progress.**

All manuscripts and synopses received by the publisher, whether solicited

or unsolicited, should be acknowledged as soon as received. The author may be told at that time when to expect to hear further, but in the absence of any such indication at least a progress report should be sent by the publisher to the author within six weeks of receipt. A longer time may be required in the case of certain works – e.g. those requiring a fully detailed assessment, particularly in cases where the opinion of specialist readers may not be readily available, and in planned co-editions – but the author should be informed of a likely date when a report may be expected.
Note: It is important for the publisher to know if the manuscript or synopsis is being simultaneously submitted to any other publisher.

7. **The publisher must not cancel a contract without good and proper reason.**

It is not easy to define objectively what constitutes unsuitability for publication of a commissioned manuscript or proper cause for the cancellation of a contract, since these may depend on a variety of circumstances. In any such case, however, the publisher must give the author sufficiently detailed reasons for rejection.

When the publisher requires changes in a commissioned manuscript as a condition of publication, these should be clearly set out in writing.
Note: In the case of unsolicited manuscripts or synopses, the publisher is under no obligation to give reasons for rejection, and is entitled to ask the author for return postage.

Time

If an author fails to deliver a completed manuscript according to the contract or within the contracted period, the publisher may be entitled (inter alia) to a refund of monies advanced on account. However, it is commonly accepted that (except where time is of the essence) monies advanced are not reclaimable until the publisher has given proper notice of intent to cancel the contract within a reasonable period from the date of such notice. Where the advance is not reclaimed after the period of notice has expired, it is reasonable for the publisher to retain an option to publish the work.

Standard and Quality

If an author has produced the work in good faith and with proper care, in accordance with the terms of the contract, but the publisher decides not to publish on the grounds of quality, the publisher should not expect to reclaim on cancellation that part of any advance that has already been paid to the author. If, by contrast, the work has not been produced in good faith and with proper care, or the work does not conform to what has been commissioned, the publisher may be able to reclaim the advance.

Defamation and Illegality

The publisher is under no obligation to publish a work that there is reason to believe is defamatory or otherwise illegal.

Change of Circumstance

A change in the publisher's circumstances or policies is not a sufficient reason for declining to publish a commissioned work without compensation.

Compensation
Depending on the grounds for rejection,
(i) the publisher may be liable for further advances due and an additional sum may be agreed to compensate the author, or
(ii) the author may be liable to repay the advances received.

In the former case, the agreement for compensation may include an obligation on the author to return advances and compensation paid (or part of them) if the work is subsequently placed elsewhere.

Resolution of Disputes
Ideally, terms will be agreed privately between the parties, but in cases of dispute the matter should be put to a mutually agreed informal procedure, or if this cannot be agreed, to arbitration or normal legal procedures.

8. **The contract must set out the anticipated timetable for publication.**

The formal contract must make clear the timescale within which the author undertakes to deliver the complete manuscript, and within which the publisher undertakes to publish it. It should be recognised that in particular cases there may be valid reasons for diverging from these stated times, or for not determining strict timescales, and each party should be willing to submit detailed reasons for the agreement of the other party, if these should occur.

9. **The publisher should be willing to share precautions against legal risks not arising from carelessness by the author.**

For example:–

Libel
While it remains the primary responsibility of the author to ensure that the work is not libellous – and particularly that it cannot be arraigned as a malicious libel – the publisher may also be liable. Libel therefore demands the closest co-operation between authors and publishers, in particular in sharing the costs of reading for libel and of any insurance considered to be desirable by the parties.

10. **The publisher should consider assisting the author by funding additional costs involved in preparing the work for publication.**

If under the contract the author is liable to pay for supporting materials, e.g. for permissions to use other copyright material, for the making and use of illustrations and maps, for costs of indexing, etc., the publisher may be willing to fund such expenses, to an agreed ceiling, that could reasonably be recovered against any such monies as may subsequently become due to the author.

11. **The publisher must ensure that the author receives a regular and clear account of sales made and monies due.**

The period during which sales are to be accounted for should be defined in the contract and should be followed, after a period also to be laid down in the contract, by a royalty statement and a remittance of monies due. Publishers should always observe these dates and obligations scrupulously. Accounts should be rendered at least annually, and in the first year of publication the author may reasonably expect an intermediate statement and settlement. The initial pattern of sales of some educational books, however, may make such intermediate payment impracticable.

The current model royalty statement (1979) issued by the Association of Authors' Agents, the Publishers Association, the Society of Authors and the Writers' Guild, or the information suggested by it, should be used as a guide, and details of the statement should be adequately explained.

The publisher should pay the author on request the appropriate share of any substantial advances received from major sub-licensing agreements by the end of the month following the month of receipt (providing monies already advanced have been earned, and proper allowance made for returned stock; allowance may also need to be made if very substantial advances have been outstanding for an extended period of time).

The publisher should be prepared, on request, to disclose details of the number of copies printed, on condition that the author (and the agent) agree not to disclose the information to any other party.

Publishers should be prepared to give authors indications of sales to date, which must be realistic bearing in mind either unsold stock which may be returned by booksellers or stock supplied on consignment.

12. **The publisher must ensure that the author can clearly ascertain how any payments due from sub-licensed agreements will be calculated.**

Agreements under which the calculation of the author's share of any earnings is dependent on the publisher's allocation of direct costs and overheads can result in dissatisfaction unless the system of accounting is clearly defined.

13. **The publisher should keep the author informed of important design, promotion, marketing and sub-licensing decisions.**

Under the contract, final responsibility for decisions on the design, promotion and marketing of a book is normally vested in the publisher. Nevertheless, the fullest reasonable consultation with the author on such matters is generally desirable, both as a courtesy and in the interests of the success of the book itself. In particular the author should, if interested and available, be consulted about the proposed jacket, jacket copy and major promotional and review activities, be informed in advance of publication date, and receive advance copies by that date. When time permits, the publisher should consult the author about the disposition of major sub-leases, and let the author have a copy of the agreement on request.

14. **The integrity of the author's work should always be protected.**

The author is entitled to ensure that the editorial integrity of the work is maintained. No significant alterations to the work (i.e. alterations other than those which could not reasonably be objected to) should be made without the author's consent, particularly where the author has retained the copyright.

The author who has retained ownership of the copyright is entitled also to be credited with the authorship of the work, and to retain the ownership of the manuscript.

15. **The publisher should inform the author clearly about opportunities for amendment of the work in the course of production.**

The economics of printing make the incorporation of authors' textual revisions after the book has been set extremely expensive. Publishers should always make it clear to authors, before a manuscript is put in hand, whether proofs are to be provided or not, on whom the responsibility for reading them rests and what scale of author's revisions would be acceptable to the publisher. If proofs are not being provided, the author should have the right to make final corrections to the copy-edited typescript, and the publisher should take responsibility for accurately reproducing this corrected text in type.

16. **It is essential that both the publisher and the author have a clear common understanding of the significance attaching to the option clause in a publishing contract.**

The option on an author's work can be of great importance to both parties. Options should be carefully negotiated, and the obligations that they impose should be clearly stated and understood on both sides. Option clauses covering more than one work may be undesirable, and should only be entered into with particular care.

17. **The publisher should recognise that the remaindering of stock may effectively end the author's expectation of earnings.**

Before a title is remaindered, the publisher should inform the author and offer all or part of the stock to the author on the terms expected from the remainder dealer. Whether any royalty, related to the price received on such sales, should be paid is a matter to be determined by the publisher and the author at the time of the contract.

18. **The publisher should endeavour to keep the author informed of changes in the ownership of the publishing rights and of any changes in the imprint under which a work appears.**

Most publishers will expect to sign their contracts on behalf of their successors and assigns, just as most authors will sign on behalf of their executors, administrators and assigns. But if changes in rights ownership or of publishing imprint subsequently occur, a publisher should certainly inform and, if at all possible, accommodate an author in these new circumstances.

19. **The publisher should be willing to help the author and the author's estate in the administration of literary affairs.**

For example, the publisher should agree to act as an expert witness in questions relating to the valuation of a literary estate.

20. **Above all, the publisher must recognise the importance of co-operation with the author in an enterprise in which both are essential. This relationship can be fulfilled only in an atmosphere of confidence, in which authors get the fullest possible credit for their work and achievements.**

Example of request for marketing information

TEXTBOOK PUBLISHERS INC.
999 WEST RIVER RD.
NEW YORK

Dr. James M. Hooper
College of Letters and Sciences
University of East Massachusetts
Ambervile, Massachusetts

Dear Dr. Hooper

We are in the process of developing the marketing plan for your book, ______________________. As the author, you are in the best position to provide us with the information we need to effectively sell your book. The College Marketing Department is responsible for coordinating all selling and promotional activities. We will be providing information to the appropriate selling operations within Textbook Publishers Inc., and will be arranging for advertising and promotion support.

In order for us to most effectively do our job, we will need an accurate description of your book. The enclosed Marketing Questionnaire will give you an idea of the kinds of information which we need. Please complete this form in as much detail as possible, using additional pages where necessary. Whenever possible, describe how each feature of your book will benefit the student and instructor, and list specific chapters and sections that best illustrate each feature.

In order for us to prepare our final plan and begin the development of sales and advertising material, we will need your fully completed Marketing Questionnaire as soon as possible. If you have any questions, please feel free to call on me.

I look forward to hearing from you.

Sincerely,

Marketing Manager
Science and Math

TEXTBOOK PUBLISHERS INC.
COLLEGE DIVISION
MARKETING QUESTIONNAIRE
SALES INFORMATION PAGE

1. TITLE:

2. AUTHOR:
(Name and affiliation as you wish it to appear in advertising and sales literature)

3. ABOUT THE BOOK:
A. OVERALL CONTENT/STRUCTURE

B. (FOR REVISIONS—NEW/REVISED ELEMENTS)

4. DISTINGUIS ING FEATURES
1)

2)

3)

4)

5)

5. TEACHING AND LEARNING AIDS IN THE TEXT:

6. SUPPLEMENTARY TEACHING MATERIALS:
(Materials not included in the text itself, such as study guide, workbook, laboratory manual)

7. TABLE OF CONTENTS:

8. MAJOR COMPETING BOOKS:

Educational Markets for your Book

Indicate below the departments in which your book might be used and the most common course titles in each department.

Department	Course Title	Level (F-So-Jr-Sr-G)	Prerequisite Courses

Will your book be suitable for use in educational or training programs in:
( ) High Schools? ( ) Preparatory Schools? ( ) Business Schools?
( ) Technical Institutes? ( ) Industrial Training Programs?
( )Government Training Programs?

Please list the names of important firms or people to contact:

Are there any Societies or Associations that would be particularly interested in your book? Give names of the best people to contact.

Are there any other educational programs (including continuing education and adult education programs) that have courses suitable for your book? Give names of people to contact.

Advertising and Sales Promotion

1. What are the three most important academic or professional journals? List in order of importance:

a)

b)

c)

2. Select several illustrations (photos, charts, diagrams, etc.) from the book that would be especially suitable for use in advertising.

3. Provide a list of the most important journals to which we might send a copy of your book for review. Make this list as extensive as possible. If you have a close association with a journal as frequent contributor or editor, please so indicate next to that listing. Use an extra sheet if necessary.

4. Provide the name and address of the news office of your university or company. List any other contacts or publications in which reviews of your book might be expected. Include organizations which you may have been associated with in the past.

5. If you lecture outside of the classroom on the subject covered by your book, and promotion of your book might be coupled with your lectures, please provide details.

6. Please make any other comments or suggestions regarding the marketing campaign for you book.

Markets for your Book other than Educational Institutions

List on this page any possibilities for direct sale of your book to professional or other in-service book buyers

Would your book be of value as a handbook or reference for professionals?

For what reasons might professionals buy your book? What practical information, tables, formulas, bibliographies, or other things could the professional apply directly to work-related situations?

List some specific job titles of people most likely to use your book: (for instance, sales manager, music teacher, chief engineer).

Where might we obtain mailing lists of potential professional users? Directories of specialized groups, magazine subscribers, society membership lists, or other lists of interested groups? If you can furnish us with any of these lists, please so indicate.

International Markets

Use this page if your book has potential international sales.

Listed here in decreasing order of importance are the key international market areas for Textbook Publishers Inc: Europe, Australia, British Commonwealth, Middle East, Scandinavia, Latin America, Netherlands, Japan, Africa, South East Asia.

Answer the following questions with the important markets in mind.

1. List the foreign countries in which you feel your book will sell best. Why?

2. List specific universities outside North America which might be receptive to your book. Include the names of personal contacts.

3. Cite examples in your book which indicate its global orientation. Use chapter or section numbers.

4. Have you cleared all permissions for worldwide rights in English?

5. Have you cleared permission for sale of translation rights for all material?

6. Have any of your previous publications been translated? If so, indicate the language(s) and publisher(s).

PUBLICATION PROGRESS
FROM MANUSCRIPT TO FINISHED BOOK

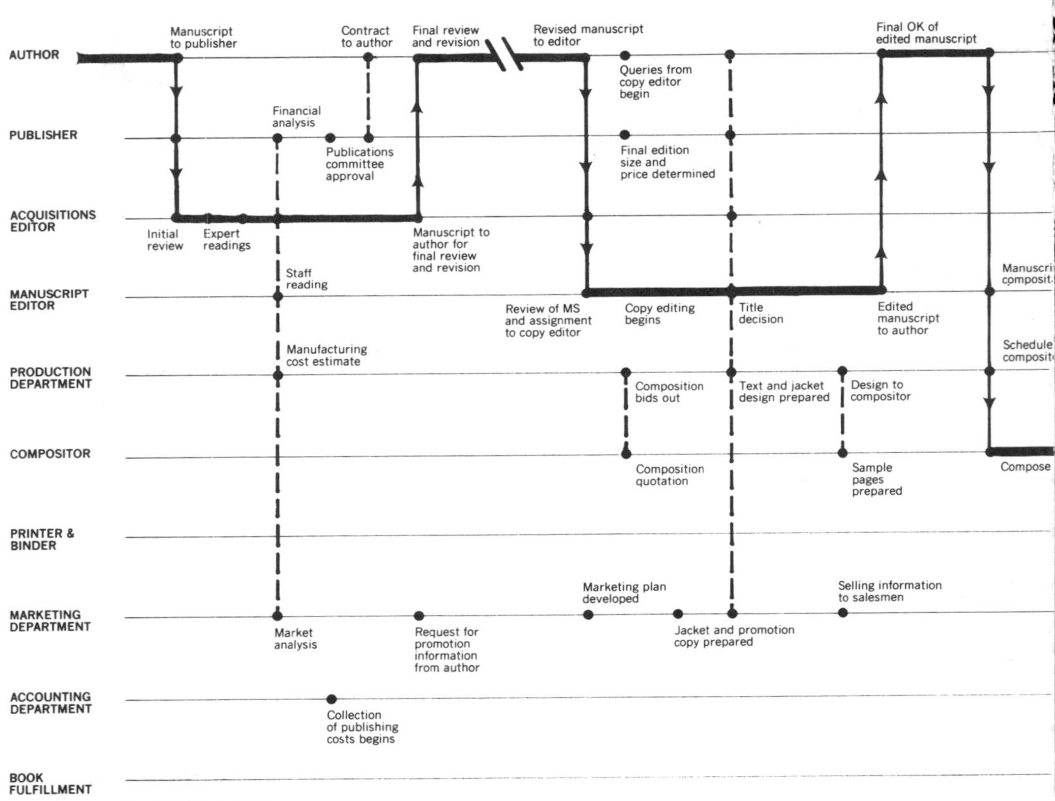

Designed by S. Diman